Einstein and Human Consciousness

Eternity is an Instant

By

Brad Buettner

iUniverse, Inc.
New York   Bloomington

iUniverse books may be ordered through booksellers or by contacting:

iUniverse
1663 Liberty Drive
Bloomington, IN 47403
www.iuniverse.com
1-800-Authors (1-800-288-4677)

ISBN: 978-0-595-52191-3 (pbk)
ISBN: 978-0-595-50944-7(cloth)
ISBN: 978-0-595-62254-2 (ebk)

Printed in the United States of America

iUniverse rev. date: 11/21/2008

A Nimble, Unconventional Nuance of Consciousness— NUNC

For Mary

Contents

Introduction

Einstein and Human Consciousness addresses a basic division in human outlook: the practical versus the abstract. Although physics, particularly the relativity of time, provides the foundation for the book's thesis, the emphasis is on people not physics. Moreover, this book is for the general reader who has little, if any, knowledge of physics.

To couple human consciousness with physics requires a firm foundation, where terms are well defined and understood. Accordingly, Part 1 begins the process, defining total reality as more than the reality that our senses perceive directly. Normally, mankind only considers the reality that came After The Big Bang as applicable to human activity. However, that reality had a source, and any consideration of all reality must include that source. That is, the identification of reality that can affect humanity must include **both** After The Big Bang reality **and** Before The Big Bang reality. That view can be hard to accept, of course, since it implies that we interact with a reality beyond that which our senses detect and which has long disappeared. Another problem is that discussion of any alternative form of reality can lead to mysticism, magic, or mystery. This work has ground rules that eliminate the latter issue. It requires that all conclusions must be measurable and observable by the senses to be judged valid, even though the senses may only perceive the result, not the process.

Part 2 is the longest and most challenging part of the book, since it initiates a search for the unlikely possibility of human interaction with Before The Big Bang reality. Because the likelihood is so remote, all terms must be clearly understood. Consequently the characteristics of any such interaction are first labeled and then precisely defined. This relatively lengthy process is well worth the reader's effort, since it establishes a firm foundation to search for Before The Big Bang [BTBB] interaction. Interestingly once the foundation is set, that interaction is rather easily found, although not in a pure form. This limitation shouldn't be surprising, though, since BTBB reality has such a radically different composition than After The Bang reality.

Part 3 analyzes patterns of human behavior and couples them with interaction with total reality, that is, with reality both Before and After The Bang. Those who choose to mainly interact with After The Bang reality are referred to as Doers, and those who enjoy delving into BTBB reality are Dreamers. However, all humans regularly interact with both realities, albeit imperfectly when dealing with BTBB reality.

Within the treatise there is nothing new presented as far as physics is concerned. The physics has been experimentally proven and is accepted, hundred-year-old information. However, the application of this old information is original.

Again, this book is for the general reader, and the physics doesn't require an in-depth understanding. Nevertheless, time and why time varies is a critical component, since it provides the basis for observation and measurement of a reality that includes everything that human consciousness perceives. The result is precise, leaving little room for subjective interpretation.

The author's formal education includes a B.S. in physics from Benedictine University (as an undergraduate, secretary of the local chapter of $\Sigma \Pi \Sigma$, the national physics honor society) and an M.S. in metallurgy from Lehigh University. Although early twentieth-century physics is used as the work's foundation, the main body of the text is based on the author's personal experience in dealing with people during a long tenure in engineering and management. This career included over 24 years of wide-ranging assignments, from introducing radical new technology, to managing cost structure, to administration of a union shop of several hundred workers. The references within the work are suggested to provide an interested reader with a deeper understanding of the physics involved, although such a pursuit is not necessary to understand the premise or conclusions presented. Internal references include Brian Greene's *The Fabric of the* Cosmos, Alfred A. Knopf 2003, Alan Lightman's *Great Ideas in Physics*, McGraw-Hill 2000, and Paul Davies' *About Time*, Touchstone 1996. Other suggested readings are:

E = mc2 by David Bodanis, Barkley 2000
Black Holes and Time Warps by Kip S. Thorne, W.W. Horton 1994
E=Einstein by Donald Goldsmith and Marcia Bartusiak, Sterling Publishing 2006
A Brief History if Time by Stephen W. Hawking, Bantam 1988

Part One

Defining Total Reality

Chapter 1
Personalities and Time

Do you prefer daydreaming to the mundane tasks of earning a living? Or do you find people who have their minds elsewhere aggravating? Do you dawdle and allow distractions to keep you from accomplishing things? Or do you set goals and meet deadlines? What's more important to you, a good song or a good buy? Are you a Dreamer or a Doer?

The conflict between personality types has plagued humanity since society began. Whether the source of the difference is generational, spousal, political or religious, circumstance often throws opposites together, and the result is a collision of values that can lead to unfortunate results. A young woman runs off with a jobless idealist, disobeying a protective parent and ending up with two hungry kids, no money, and no husband. A wife tires of a husband who spends all of his time earning money while ignoring his family, so she files for divorce. A Democrat dreams of healthcare for all, but a Republican worries that that option will wreck the economy, so healthcare languishes in a hodgepodge of uncertainty. A religious leader condemns another belief, and followers go to war rather than allow both beliefs to flourish. The result is violence, upheaval, and carnage.

There are other differences, of course, that have a resolution. For example, a hard-driving entrepreneur may fall in love with a quiet, romantic woman who sings like a bird, smiles like an angel, and writes sonnets with the ease of a summer breeze. A few years into the marriage the entrepreneur demands that his wife be more practical, while she demands that he familiarize himself with cultural pursuits. To make the marriage work they both adjust. Or an ambitious supervisor finds lackadaisical employees far too relaxed to accomplish what needs to be done. To resolve the issue the supervisor instills additional passion into the workplace, the workers respond, and their talents blossom.

But why do these differences exist in the first place? Why do some people plan and plan and plan and never get anything accomplished, while others jump into action, far more interested in the pursuit of a goal than figuring out the best way to attain it? Why do some people love poetry and others hate it? Why are there people who can't hold a job and others who become respected supervisors in their thirties? Why are there bums and alcoholics as well as workaholics and teetotalers?

There is an answer to these questions, and this book will attempt to provide that answer. Whether you're a Dreamer or a Doer, or—like most—somewhere in between, you'll learn to understand why you are like you are and why others are like they are. The answer is fundamental and has to do with two states of consciousness that we all possess. To define these states we must investigate a fundamental aspect of human life. We must investigate time.

Time is a basic qualifier for human judgment. It indicates duration, and mankind uses duration for recording life. We use time to determine payment for work and calculate interest on debt. We consider the age of an artifact in any appraisal. Time is an essential dimension in human terms, every bit as important as length, width, and depth. So to look at total reality, a reality that encompasses everything, time plays a role at least equal to the spatial dimensions.

Note that this is a measurable consideration. There's nothing mystical or magical about it. The total reality spoken of here is the total observable reality, the reality that we can verify with experimentation. It is a measurable time difference that plays the critical role in determining a Dreamer or Doer outlook. That is,

Dreamers have a different view of reality than Doers, and the reason is that Dreamers concentrate on a different reality altogether. Dreamers have found a peculiar aspect of human consciousness that has different properties than the physical reality that our senses detect. Nevertheless, this aspect of human consciousness is as measurable and actual as anything physical. To detect that other reality, however, requires a thorough understanding of the concept of time.

The reason time is so important is that it varies, at least as far as the human consciousness is concerned. That is, individuals perceive time differently depending upon the circumstances. This fundamental difference in perception hints that there may be more to total reality than we commonly think. Therefore, to confirm that possibility, time and all its implications must be scrupulously analyzed.

Time has fascinated mankind for eons. Swiss clockmakers with their mechanical marvels believed they had mastered time as they honed the mechanics of establishing a rhythm that would tell with a glance what the exact time was, that is—as they understood time—exactly where the sun was with respect to a 24-hour day. Humans have worked tirelessly at making that measurement even more precise, advancing the art to the use of atomic clocks that measure time to infinitesimal fractions of a second. When conspirators synchronize their watches these days, they get it right.

But understanding time involves much more than improving the accuracy of measuring duration. Intellectuals have long agreed that time goes beyond the 24-hour daily cycle. There are cultural differences as far as time is concerned. For instance, one society may consider being on time for an appointment in terms of plus or minus days while another, such as ours, considers punctuality in terms of plus or minus minutes.

The different perceptions of time extend even beyond the cultural. At the beginning of the twentieth century, Albert Einstein discovered a relationship between velocity and time. He'd been asking himself questions on how the universe works, and he rightly concluded that the velocity of light limits the velocity with which information can travel. With that limitation and the premise that light cannot go faster than its defined velocity even if the source is moving, he reached the inescapable conclusion that time varies depending upon the relative velocity of the observers. If they have the same velocity, they agree on the time, but if they are moving with respect to each other, they don't. This conclusion, as astounding as it appears, has been verified experimentally, so challenging it is like maintaining that the earth is flat or that the sun moves around the earth and not vice versa—one can make the challenge but cannot win it and claim to be reasonable as well. In the next chapter we'll demonstrate this mathematically. Don't be scared off. The demonstration is surprisingly simple, and the fact that it is lends even more credence to the concept. In these days when eminent physicists flounder when considering the mathematics of String Theory, the simplicity of the mathematics of Einstein's Special Theory of Relativity is indeed refreshing.

First, though, let me emphasize that this book isn't about physics. Though trained as an undergraduate in that discipline, I claim no such expertise. This is a book about time and how its interpretation affects our outlook on reality. This is a book about **NOW**. As Brian Greene says in *The Fabric of the Cosmos*, "*...In this way of thinking, events, regardless of when they happen from any particular perspective, just are. They all exist. They eternally occupy their particular point in spacetime.*" That is, the NOWs all exist, and they all exist right now. Not to us, of course, but once they blossom, they will stay in spacetime forever. That's what this book is about.

There's a huge jump in logic here, and the implications are mind-boggling. We need to back up and go at this more slowly. I caution you, though. Don't expect some off-the-wall conclusions later on, because there aren't any, despite the above.

Okay, now open your mind a bit, so that you can accept that time is relative to velocity, and we'll begin. (More precisely, time is relative to the physical circumstance. That is, the difference in rates of time depends not only on motion but on spatial and mass considerations. For our purposes, however, we'll keep it simple and refer only to velocity.) Once you appreciate time more completely, you'll be able to apply that understanding in finding out how consciousness works.

There have been countless books written on time, and many of them involve physics. Some might question what time has to do with physics, but Albert Einstein's brilliant discovery that time was relative began a revolution in the early 1900s that forever changed the way scientists think about time in the physical universe. It's in that revolution that we will find our understanding of consciousness.

Again, this book isn't about physics. It's about people, but to understand people, we must understand time. And to understand time, we must understand its basic principle, so in the first part of this discussion we will define how the issue of time leads to a more complete understanding of total reality. That, however, involves physics.

No one would be more surprised at linking time to consciousness than Albert Einstein, who, despite his genius, wasn't known as a people person. We'll make this foray into physics entertaining, I promise. We'll keep the dry mathematics to a minimum and provide just enough information to allow an understanding of the subject at hand. Once you see how time involves human consciousness, you may even want to pursue the physics further, but if you do, you won't have this author to guide you. Physics is best left to the physicists.

The concept of the relativity of time is surprisingly simple, requiring nothing more than early high school geometry and algebra to grasp it. If you made it to the junior year, you should be able to follow the theory. Then, once you comprehend the idea, you can work with it, and draw conclusions that will affect how you look at people, and how you look at life. In fact, the theory extends even beyond that, so it's well worth the effort to plow through these first few chapters.

Most people, when first confronted with the concept that time is relative to velocity consider the idea either as fiction or so complicated that only a genius could understand it. The proof that relativity is impossible, disbelievers may argue, is right there in a clock. A minute takes just that—60 seconds. If it took longer than

60 seconds, it wouldn't be a minute. How could velocity change that? And what has the velocity of light got to do with it?

The key to understanding the issue is to consider how we know that a minute is a minute. Something has to give us that information. Information has to flow, and the time it takes the information to flow is the heart of the issue. When we read a clock, we believe that what we've read is the time the clock says, because the velocity of light is so fast. The time it takes the information to reach us is so insignificant that the difference between what we see and what the clock actually says is nil. But what if light traveled much slower? Say, for example, that light traveled at one foot a minute. That's impossibly slow, of course, but the example will be helpful in understanding the argument. Now if light traveled at one foot a minute and the clock was five feet away, the time you would read would always be five minutes late. When you saw one o'clock , it would be 1:05 on the clock. 1:05 would be 1:10. You'd always be five minutes off. Let's say the clock began to move away from you at one foot a minute. The new position of the hands would never reach you, because the light carrying the information wouldn't reach you. From the clock you'd conclude time was standing still even though a watch you held close to your eyes would tell you otherwise. Also, note that when the clock was still, your interpretation of the time difference was constant—5 minutes late. You agreed on the duration of a minute but disagreed on which minute it was. When it moved away from you, however, your interpretation of a minute conflicted with the clock's. The clock's minute hand kept moving, but that information never reached you. Actually, from your point of view, your time had moved, but the clock's hadn't. It had slowed to zero. If the clock was moving away at a slower velocity, your time would stay at the same pace, but the moving clock's time would appear to be slower than yours, only by not as much. Thus the Special Theory of Relativity states that something moving with respect to you has a slower time.

The above example is trivial, since it uses such a slow velocity for light, but it does demonstrate that if information doesn't flow instantaneously, time isn't constant. Time depends on the circumstance. This idea seems to contradict all we have learned since childhood, so it isn't easy to accept.

If I asked you to drop a bowling ball on your foot from waist-high, you wouldn't do it. You wouldn't need any trivial example or equation, no matter how simple it was, to tell you that you'd end up in severe pain and possibly with a broken foot. Although Newton's Law of Gravity is essentially no more involved than Einstein's Special Theory of Relativity, you've had enough experience since childhood to warn you of the danger. Not so with relativity. The velocity of light is fast, so fast that the adjustment for relativity is small, far too tiny to observe in daily life. It involves squaring the velocity of light, that is, multiplying it by itself and using that number as a divisor for the square of the velocity of the moving clock. For example, if the velocity is 60 MPH, the adjustment involves dividing 60×60 or 3600 by 448,000,000,000,000,000. So any subtraction in a time equation due to normal movement isn't going to be very much.

The difference in time due to velocity is not only imperceptible, but it wasn't even measurable when the theory was proposed. Imagine the ridicule simpler minds must have given Einstein when they first heard his proposal. Even today there are those who think that the relativity of time is some kind of gimmickry meant to fool the less sophisticated. Yet the mathematics isn't that complicated nor is the logic mind-boggling. The one premise—that light can only travel at one velocity—could pose a problem, but the idea has been tested and proved over and over. At present there is no excuse for not believing it. Still, understanding the potential of a falling bowling ball comes much easier than understanding the implications of time.

Since the consequences of time relativity are extremely minute, their measurement offers little help. Atomic clocks run slower in jets but the fraction of a second difference is tiny even after a long flight. Time relative to velocity means that returning astronauts haven't aged as much as the people left behind, but again, the difference isn't enough to notice—or even to combat the muscle atrophy that occurs outside gravity. The rest of us, who may take an occasional jet flight but travel mostly at automotive speeds, don't see any result at all, so there's no reason to consider time relativity as a solution to Ponce de Leon's quest.

The problem with written communication is that unlike the theater there is no immediate feedback. I have no idea whether readers with many different backgrounds have agreed at this stage to accept the relativity of time. If you're one who still hasn't, there are many other books on the subject, and, with a little more reading, you'll be convinced that time is just as relative as weight. (Yes, that bowling ball's weight is relative to mass. On the moon it would weigh only a fraction of what it does on earth. In space it weighs nothing. So you see, that instinct you had with my earlier proposal of dropping it from waist-high wasn't all that correct either.)

It's not important that you know all the nuances of the time factor, but it is important that you agree that time is relative to circumstance, and that if I accelerate and go faster than you, my time will be slower than yours. If you can cling to that notion with the same sincere belief that you cling to the notion of a dropped bowling ball hurting your foot, you'll be able to follow the arguments of this book, arguments that flow logically from this proven view of reality. Again, velocity **slows** time. Believe it. Accept it. There's too much evidence out there to bother with any skepticism. **Velocity slows time!**

Want more convincing? Okay, but brace yourself because it requires some math. We'll have some fun in the next chapter. Don't skip it, because if you do, you'll be wondering how scientists are so sure of themselves. The mathematics is conclusive, just as conclusive as one plus one equaling two. Granted, the logic is a bit more involved, but the mathematics isn't. So come along. We'll have Bill and Steve bolt their wagons together and make a couple of measurements.

Chapter 2
A Mathematical Demonstration of Time Relativity

If you're used to a mathematical approach, you'll find this one easy to follow, but if you're not, you're in for an even more enjoyable experience. The approach will not only convince you of the relativity of time, but it will give you confidence in a mathematical view as well.

First, let's review the properties of a right triangle. For those long out of school, it's a triangle where two sides meet at a 90° angle. The third side, the one opposite the 90° angle, is called the hypotenuse. If you recall from simple geometry, the square of the length of the hypotenuse equals the sum of the squares of the length of the other two sides. That is, $a^2 + b^2 = c^2$. With that in mind, let's set up an experiment.

Bill and Steve are brothers, and, like most brothers, they're not quite sure whether the other one is trying to pull a fast one. So when Steve proposes to measure the time it takes for light to reach a mirror and bounce back, Bill is skeptical, but he agrees to go along. They rig up a bulb and mirror and prove to themselves that the distance between the bulb and mirror is constant whether the contraption is moving or not. Then they mount the contraption on Bill's wagon and bolt Steve's wagon alongside, so that someone can travel in one of the wagons at the same velocity as the light and mirror. Figure 1 shows the result.

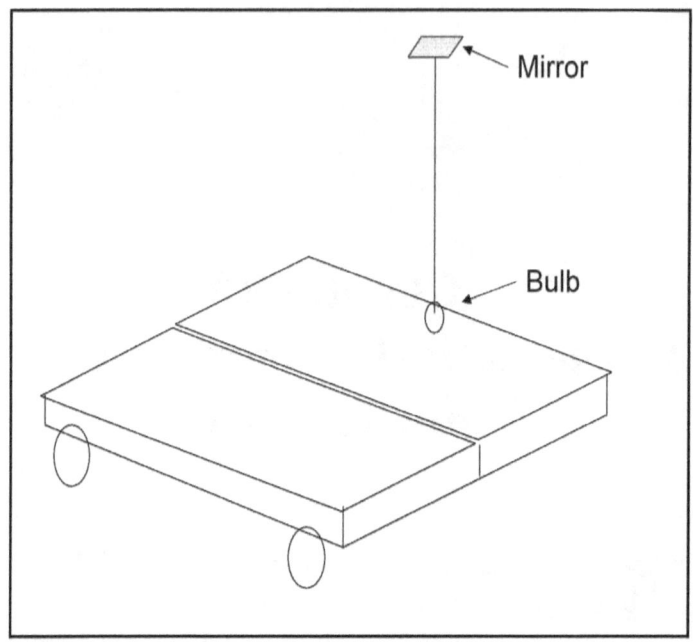

Figure 1

The brothers recruit Jason, the neighbor across the street, to pull the pair of wagons at a constant velocity we'll call v. Steve will ride in his wagon while making his measurement, but Bill will make his observation watching the wagons pass. They both are so good with video games that they can use their fast fingers and precise stop watches to time exactly how long it takes for the light to reach the mirror and return to the bulb. (Of course, this is impractical, but with detectors and an accurate timer it could be done.) Since Steve is moving, and Bill is standing alongside, they will have a slightly different view of the event. Note figure 2.

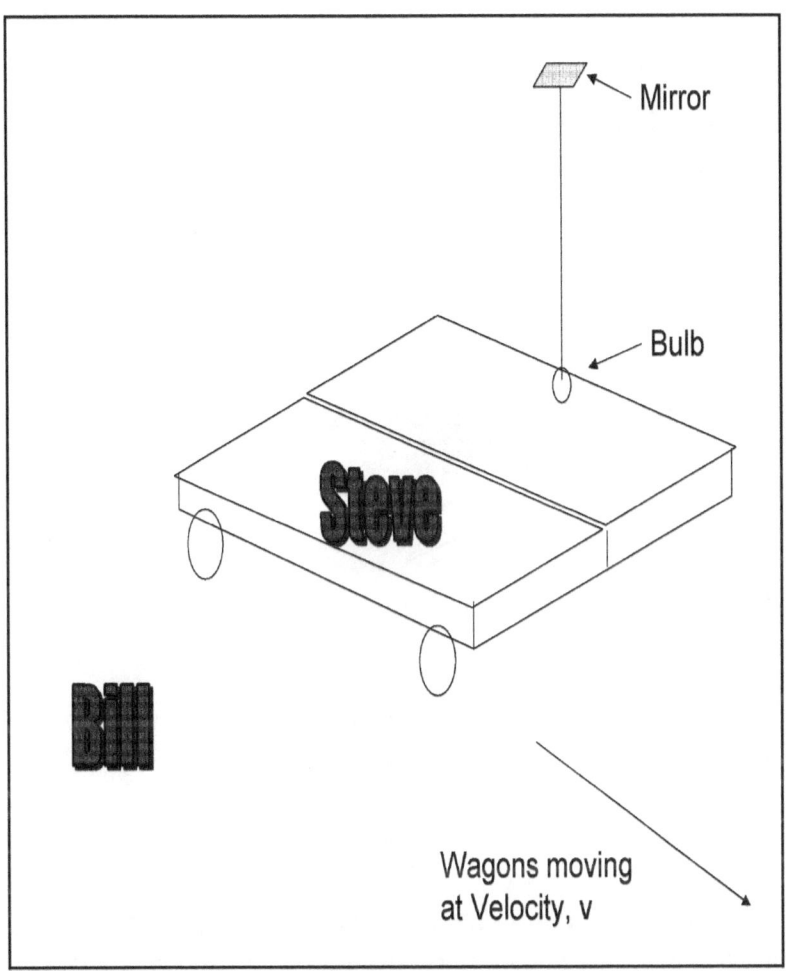

Figure 2

Steve sees the path for the light going to the mirror and back as perfectly vertical as shown in Figure 3.

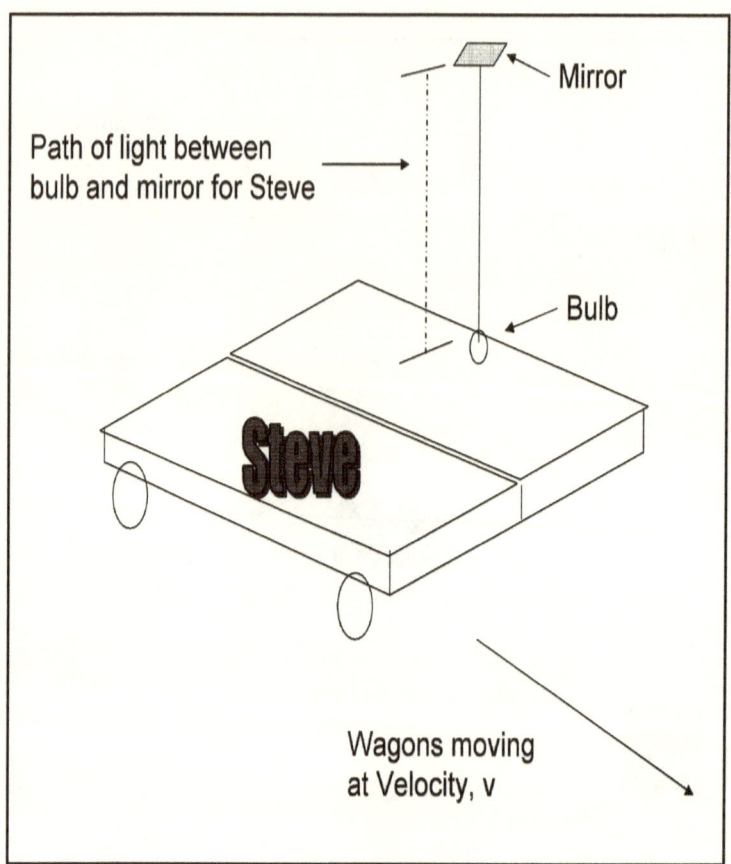

Figure 3

Since Bill is watching the wagons pass, the light at the bulb will be moving with the wagon and, therefore, from his perspective the light will travel to the mirror at an angle and return in the same manner. This is shown in Figure 4.

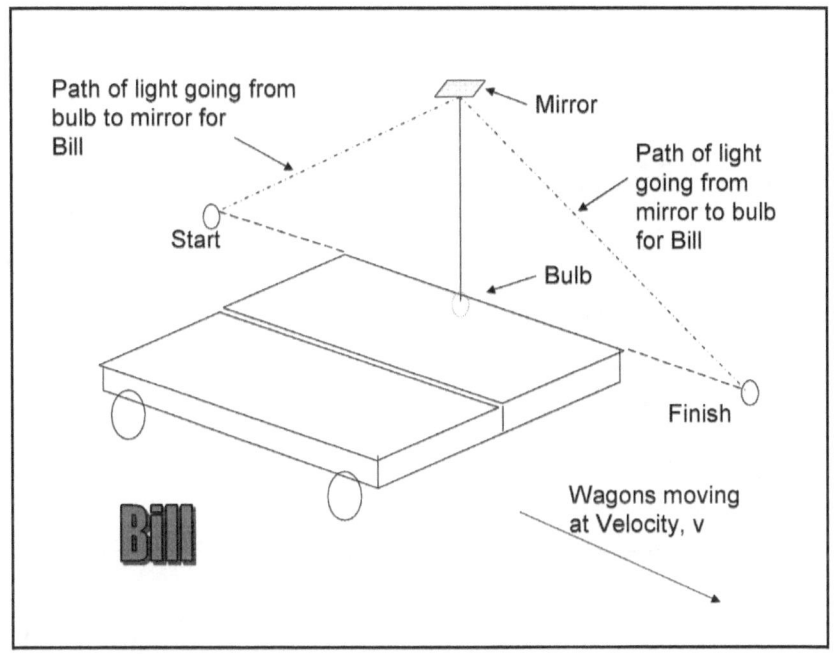

Figure 4

Let's look closely the path for the light from Bill's perspective. We'll call the time he measures for a complete cycle of light going to the mirror and returning as T_b. We'll call the distance between the mirror and the bulb D. The velocity of light is c. So, in figure 5 we have the path of the light as Bill sees it. The light travels from the bulb at the start point to the mirror at the midpoint and back to the bulb at the finish while the wagons pass.

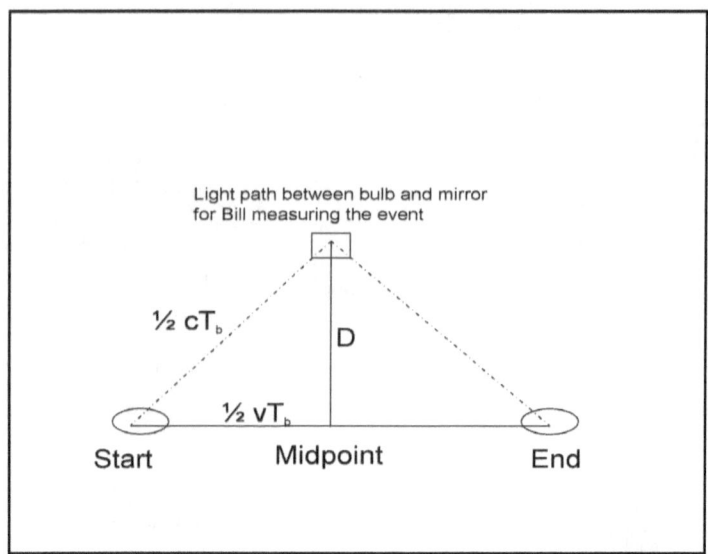

Figure 5

The first half of the trip can be described by a right triangle. One side, the distance from Start to the Midpoint of the event, can be equated to the wagons' velocity multiplied by half the time Bill measures for the event, $\frac{1}{2}vT_b$. Another side is D, the agreed distance that the mirror is from the light bulb. The hypotenuse of this right triangle is the light traveling at the angle as it goes from the bulb to the mirror, again $\frac{1}{2}$ the time, since we're only looking at one way, multiplied by the velocity of light.

Remember $a^2 + b^2 = c^2$? This is the Pythagorean Theorem mentioned earlier. Let's apply it here.

$$D^2 + [1/2 \, vT_b]^2 = [1/2cT_b]^2$$

But D can also be expressed as $\frac{1}{2} cT_s$, the velocity of light times half of Steve's time measurement. Remember, for this experiment, Steve is moving and Bill is stationary. Bill's time is T_b and Steve's time is T_s.

So if D is replaced by $\frac{1}{2} cT_s$, the equation becomes

$$[\tfrac{1}{2} \, cT_s]^2 + [\tfrac{1}{2}vT_b]^2 = [\tfrac{1}{2}cT_b]^2$$

Solving the equation for T_s—remember T_s is for the moving observer, Steve—we get:

$$T_s = T_b\sqrt{1 - v^2 / c^2}$$

(If you can't remember how to do this, it's not much of a chore. I'm simply combining terms and isolating the two measurements of time. For those interested in the algebra:

$[½ c T_s]^2 + [½vT_b]^2 = [½cT_b]^2$

or $[½ cT_s]^2 = [½cT_b]^2 - [½vT_b]^2$

or $T_s^2[1/2c]^2 = T_b^2[(1/2c)^2 - (1/2v)^2]$

or $T_s^2 = T_b^2[1 - v^2/c^2]$

Taking the square root gives the final equation.)

Let's take a close look at this equation and note exactly what's in it. The equation includes the time of a moving observer, T_s, and the time of a stationary observer, T_b, for measuring the same event—the time for the light to leave the bulb and get reflected back from a mirror. It also includes the velocity of the moving wagon, the velocity of light, the number 1, and a square root sign. Note also that since v^2/c^2 is a squared number, it's always positive. The result will be subtracted from one. Since v cannot ever be greater or equal to c—only light can go the velocity of light and nothing can go faster—the number under the square root will always be less than one but more than zero. Thus, the moving observer, Steve, always measures time as a fraction of that of the stationary observer, Bill. That is, Steve's clock is slower than Bill's, since Steve measures the event as taking less time. Remember, both Bill and Steve measured the same event. The only difference was that Steve was moving with respect to Bill.

Significantly, the difference **is** dependant on the wagons' velocity. The faster the wagons go, the slower Steve's clock is compared to Bill's for measuring the same event. That is, the faster Steve travels, the smaller the number under the square root sign gets. As that number gets smaller, T_s becomes smaller when compared to T_b. Or—and this is what's critical—**velocity slows time down!**

Again, there's nothing new here. This result is a hundred years old. I've cut a few corners, so you may wish a more detailed explanation. A better treatment of the subject is in Alan Lightman's *Great Ideas in Physics*. He goes into the concept at length and yet still keeps the mathematics to a minimum. The important fundamental concept that needs to be ingrained in your mind, however, is that velocity slows time.

At this point you may be asking yourself a question. Say, for instance, that I'm moving relative to you. That means my time is slower than yours. However, from your viewpoint you're moving relative to me, so your time must be slower than mine. If it's all relative, who's to say which one of us is the one that's moving, you or me? It would seem that from both points of view each time has to be slower than the other's. But that's an impossible dilemma. Both times can't be slower. So which time actually is?

This dilemma is more puzzling to me than the fact that time is relative to velocity. Its resolution is found when the two of us come in contact with each other. It's in that interaction that we discover whose time was slower. Of course, in order to come in contact, we have to communicate, and the circumstances of that communication dictates who experiences the slower time. For instance, consider space travelers coming back to earth after an almost-the-speed-of-light journey. The astronauts will have aged less than those of us who were earthbound. But the way we draw that conclusion is by measuring the comparative ages after the voyage. That comparison defines the circumstance of the communication. They began the journey in our timeframe and ended it in that same timeframe. Thus, their clocks were slower. If in some odd way, we could have gotten earth to move away from the rocket ship and then brought it back again, our clocks would have been the ones that slowed relative to the astronauts. The reference for the communication determines how any difference in clock speed is noted. If you'd like a more detailed explanation of this, Paul Davis gives an excellent treatment in *About Time*. This issue is very important when comparing rates of time. The difference isn't noticeable until the two timeframes intersect. So, if my timeframe is ten times faster than yours, I won't notice the difference and neither will you unless we somehow communicate. It's in that communication, that intersection of timeframes, that the discrepancy surfaces. Therefore, when thinking about time, remember two things: 1, velocity slows it down, and 2, the difference is only noticed when there's interaction between the two timeframes.

As already noted, v^2/c^2 is effectively zero for anything we do on a daily basis. Geniuses like Newton, Galileo, and da Vinci could ignore the now obvious issue of time relativity and apply their theories quite ably. It wasn't until scientists became more sophisticated in their questions about the universe that Einstein's idea became important, and then its importance was for physical measurements. So, if all the philosophers and thinkers of the past got along without considering the effect relativity of time has on human consciousness, why should it be applicable to the understanding of consciousness now?

Actually, it turns out to be basic to our existence. Understanding the implications of the relativity of time will show that there is another aspect of our consciousness, of reality, that must be considered in order to have a total picture of what reality is. Mankind hasn't considered it before in these terms only because it has used other terms instead. Usually humans involve some sort of mysticism to offer an explanation of total reality. These mystical terms are merely a fog in a poor attempt to explain our consciousness. Now that fog can be lifted, and we can look at reality with fresh and enlightened eyes.

Chapter 3
The NOWS of the Universe

What does the fact that velocity slows time have to do with consciousness? To answer that question, let's take a hard look at what we consider reality. To experience the world around us we taste, touch, smell, see, and hear, and with these senses we make judgments about reality. If, for some reason, we misinterpret what our senses tell us, someone will correct us soon enough and convince us to change our minds. That occurs often enough, particularly in our formative years.

On the other hand, what if everyone's wrong? What if everybody is interpreting reality incorrectly? It happens, of course. Eminent thinkers have pronounced a particular reality with certainty only to be proven incorrect later on. Long ago Copernicus pointed out a different way of looking at the heavens. When Galileo agreed, the establishment of the day—the Roman Catholic Church—considered the decision an outrage. It believed Galileo's viewpoint challenged religious beliefs and condemned that viewpoint out of hand. Now, of course, the conflict appears absurd, but it didn't then. Einstein's theories provide a more recent example of reinterpretation of true reality, but he didn't end the controversy. Arguments about what reality actually is will continue as long as mankind theorizes about how the universe works.

Besides scientific arguments, there are other incorrect interpretations of reality that are more subtle and widespread. The night sky is a good example. When we look up to the moon and see it floating peacefully in the blackened sky surrounded with stars, we can't help but feel enchanted. The surrounding constellations form a sparkling texture, and, if we're lucky, we embrace a loved one as we contemplate the heavens. The problem is that what we're looking at isn't reality, not contemporary reality, at least. Even for the moon, what we see is what it looked like moments ago, not currently. The reality we see is delayed reality. It's sort of like that clock that always gave a reading five minutes late. The delay for the moon may be trivial, but for a galaxy millions of light-years away, the delay is millions of years. In real time that galaxy's condition is quite different than the one we see. These incorrect observations don't pose a problem for us, however, because we define our observation **to be** the present, and we maintain that that galaxy millions of light-years away is just as real to us as the jet plane cutting through the sky directly above us.

The only way to eliminate a time delay with a distant object would be to close the gap by traveling closer to whatever we wish to observe. You may argue that this would be impossible for something millions of light-years away, since it takes light millions of years for the journey and nothing travels faster than the velocity of light. There is a way, however. If we used a very fast spaceship that could approach the velocity of light, our clocks would slow down just as atomic clocks do in a jet plane, and, if the spaceship went fast enough, the clocks would slow so much that what we considered millions of years in earth-time would be only a few years in our new time. Thus, we could arrive and see the galaxy as it really is. New stars may have been born since we first observed the galaxy from earth. Perhaps other stars would have morphed into super-novas. Still others could have collided with adjacent stars, and we would be astounded to discover that the catastrophic result was about to swallow our space ship. In any case, the reality we would observe would be quite different than the one we saw from earth.

Speaking of earth, we'd look back on it, and it would appear only a bit more than a few years older than it was when we left. Yet on earth millions of years would have gone by. The only communication we would have with the earth is what was sent right after we left and was just catching up. The messages would be from people we knew, but the messages would be old and obsolete. Those who watched us take off would be long dead as well as their cultures and even perhaps the earth itself.

Earth and our rocket ship would be greatly separated in spacetime. Earth's NOW would be different from our NOW, just as the distant galaxy's NOW is far different from ours when we look at it from earth. These different NOWs hint at the reason that time is so critical in understanding our total reality.

Supposedly we know what now is, just as we know what reality is. Now is the present, our present. On the other hand, it isn't everybody else's present. For another entity it could be the future. It could even be the past. I'm not suggesting that there is another entity out there, but I am suggesting that if there were, that entity might have a different NOW from ours.

Of all the possible NOWs, however, one of the most important is the one that got the universe going in the first place. That's the NOW that physicists before and after Einstein have long sought to define. A breakthrough in that regard came in 1929 when Edwin Hubble discovered the expansion of the universe, and the Big Bang theory of the origin of the universe took hold. Physicists have worked with that idea ever since, sharpening their pencils while theorizing about the Big Bang. They've postulated its cause and theorized about what happened microseconds later, minutes later, and years later.

What about that NOW? Is its timeframe—its speed of time— different than ours? We're moving with respect to that NOW, moving at great speed since the explosion approached the speed of light. All of matter that we know is moving with respect to that NOW, so reality as we know it is defined by a NOW that began at the Big Bang. The timeframe of all of our reality that started then is the same. That is, the duration of an earth-year in a galaxy millions of light-years away is equivalent to 365 of our days. Its information may reach us later, and thus our view of it is delayed, but its timeframe—again, the speed of its clocks—is the same in that galaxy as it is for us. However, that galaxy, as well as the rest of reality as we know it, moved away from moment of the Big Bang extremely fast, so the speed of time of our universe must be slower, much slower, than the timeframe that existed Before The Big Bang. Therefore, to contemplate total reality, we must consider the reality we experience everyday and the reality that existed earlier, even though, from our point of view, that reality has long disappeared.

We don't know anything about that earlier reality, but we can make comparisons based on what we do know about our everyday reality. For instance, everyday reality consists of physical matter that our senses pick up and a 24-hour timeframe that it takes for the earth to make a 360° spin on its axis. An earlier reality that existed Before The Big Bang must be quite different and unlike anything we experience After The Big Bang. In fact, this earlier reality is so remote to anything we normally experience that most would think it inapplicable to human existence. Besides, there may not be anything there to contemplate. Most individuals would claim that any residue from Before The Big Bang is, in actuality, After The Big Bang reality. That is, there can be no distinction between the two realities, since one was born from the other. Although the two are separate and distinct, Before The Big Bang reality has morphed into After The Big Bang reality, such that there is nothing left of the earlier reality. However, what if that isn't 100% true? What if there is some leftover that mankind has somehow found? It seems unlikely, of course, but perhaps it's worth considering. If that earlier reality still exists, admittedly a big **if** for most readers, and if humans can interact with it, a second big **if**, how would it appear to us during that interaction?

To answer that question let's consider what happened at the Big Bang, at least as far as interacting with all of reality, both before and after, is concerned. First, there's the BANG. **BOOM!** Our universe and our everyday timeframe begins.

Remember, as velocity increases, time slows. Also, Before The Big Bang there isn't matter as we know it. After The Bang matter appears and with it a slower time when compared to Before The Big Bang. How much slower, we can't know exactly, but initial speeds approached the speed of light, so a proper assumption would be that, in comparison, our speed of time is close to the speed of time of the super-fast spaceship heading to the millions-of-light-years-away galaxy. Don't forget that to make any such comparison there has to be interaction between the two timeframes. That is, astronauts have the same time as we do on earth as far as they're concerned while they're traveling in space. It's only when the ship gets back to earth that any age difference can be observed. Thus, making a comparison with the reality that existed Before The Big Bang, requires actual interaction. If that interaction occurred, our everyday timeframe would be much, much slower than the Before The Big Bang timeframe.

Back to the new matter. Perhaps it's only a tiny bit, but whatever it is, the matter eventually aligns itself and becomes the first stage of After The Bang. It's not important for this discussion as to how the matter aligns or even what the matter is. All that counts is the NOW of the new timeframe After The Bang, and the NOW of the timeframe Before The Big Bang, and what exists in those two NOWs.

The term ATB sometimes is used for After The Bang. That would be our timeframe. To differentiate the before from the after, let's call Before The Big Bang, BTBB. (I'm purposely using four letters instead of three to distinguish the after and the before more clearly.) So we have two NOWS, one BTBB and one ATB. The BTBB NOW is one of essentially no matter as we know it and a much faster time when compared to the ATB NOW. If a comparison isn't made, that is, if there is no interaction between the two NOWs, then both timeframes would proceed equally well on their own, sort of like the timeframes of earth and that fast rocket ship heading for a distant galaxy. It's only when the two NOWs interact that the discrepancy appears. When they do, however, our reality has a much slower timeframe. For want of a better term, let's label our view of this large difference in the speed of time as timelessness. This is important, and we'll devote the next several chapters defining timelessness much more clearly. For now, though, let's just say that when the two timeframes interact, ours would be the one of matter and 24-hour-a-day time, while BTBB reality is the one of nonmateriality and timelessness.

Again, whether the two NOWs interact is the crux of this topic. Do we, as humans, interact with BTBB reality?

Unlikely is a reasonable answer to that question. Also, why even try to look for such interaction? Our reality is so completely defined with ATB circumstance that to consider any other circumstance appears ridiculous.

However, the problem is that ATB reality does not fully explain all human experience, particularly when human consciousness is considered. And if ATB reality has such a limitation, there aren't a myriad of realities from which we can pick and choose. The only other reality we know of is the reality that brought our present reality into existence. Total reality consists of only two components, our present, everyday, ATB reality and any reality that preceded it. Fortunately there are observable and quantifiable differences between the two realities. So, although the possibility of any interplay with BTBB reality appears unbelievable at first, if an interaction meets carefully defined criteria, that incredulity can dissolve into a reasoned possibility. It's certainly not going to be easy to identify interactions between the two realities, and we won't be able to identify the interactions in a perfect form, since we're embedded in our material world. That doesn't mean, however, that a probe isn't worth the effort.

We'll spend most of this book analyzing human interactions, searching for hints of interaction between the two NOWs. The challenge is substantial. In ATB reality our dependence is on the physical, while in BTBB reality, matter plays a minor role, if it plays any role whatsoever. As already noted, the timeframes are far different as well, and we'll see that the timeframe differential provides the biggest clue in our search.

Like it or not, there are two basic viewpoints of reality, one from BTBB and one from ATB. When these viewpoints intersect, there is a fundamental difference in the perception of reality.

Our search for that intersection will consume Part Two of this discussion and, since the issue is so sensitive, it will be the lengthiest part. We won't get back to Doers and Dreamers until Part Three, so for that topic you'll need some patience. However, in looking for interaction between BTBB and ATB reality, we'll have some interesting insights along the way, raising issues that can only be answered by an extended view of what our total reality includes. So hop aboard for part two. Often a journey can be as exciting as the arrival.

Part Two

The Search for Interaction
with BTBB Reality

Chapter 4
Defining the First Criterion—
Timelessness

Numerous theories have been proposed to explain the Big Bang but no definitive discovery of how it actually occurred. Theories include the possibility that there have been many Big Bangs and that our universe is merely the result of the latest one. Whatever theory one chooses to believe, however, doesn't change the fundamental concept that billions of years ago in our timeframe a tumultuous event occurred. It's that event that defines the universe we're embedded in, so whatever preceded that event is the only other reality that could affect us. That's the reality that we must somehow find.

However, as already noted, to suggest contemporary interaction with the reality that preceded the event seems ridiculous, a pure fantasy. After all, the resultant explosion altered the universe so dramatically that a fair assumption is that any reality which existed before the Big Bang was irreparably altered, and, even if that assumption is wrong, finding any interaction with BTBB reality would be a seemingly impossible challenge. Nevertheless, free and open thinking allows for a fair search, even though intuitively it seems that the search would be futile.

Let's first set a few ground rules. Any interaction we find will not be proved in the sense that we actually identify BTBB reality materially. We can't do that, since matter Before The Big Bang was in such a drastically different state. On the other hand, the time difference will be clearly identifiable. That is, when the two realities intersect, the amazing difference in the speed of the clocks of the two realities will surface. It's that time differential that substantiates BTBB interaction. Unfortunately, that's the only measurable feature of the interaction. This may not be entirely satisfying, but, if the identification is significant enough, it should be sufficient for most readers.

As stated in the previous chapter, we'll discover that all interaction with BTBB reality is imperfect, because we're embedded in a universe composed of ATB material. The important point here is that the purer the interaction, the more significant it is and, therefore, the more convincing. Unfortunately, the purest form of such interaction that we'll identify is the one most of us are the least likely to experience, although we all have similar experiences to some extent.

Another expectation that some readers may have is finding a "place" where this BTBB reality resides. When we look for something and find it, we say that it exists at a certain physical point or circumstance. BTBB reality has to be different. It has to be all-encompassing and can only exist if it's pervasive in our universe. That is, the BTBB reality was fundamental to our universe. Any existence of it within the ATB universe wouldn't be localized, since our whole universe was born from that reality. Thus, we won't find a "spot" or "special circumstance" for the interaction. Instead, we'll have to infer BTBB reality in that it meets the criteria that such interaction would display.

Of course, simply demonstrating that the criteria are met isn't sufficient either. The experience must be such that only BTBB interaction can explain it. Here too, however, we run into the problem that no BTBB interaction can be pure. Thus, any experiences we do have can be explained to some degree using only ATB reality. However, ATB reality won't completely explain the experience, and that's the critical point.

Finally, some readers may feel that for a BTBB interaction to be actual it must have a physical effect as ATB interactions do. That is, interacting with ATB reality yields a physical consequence of some sort—hot surfaces burn and sharp edges cut. This doesn't apply to interactions with BTBB reality. Since BTBB reality is essentially energy, there isn't any material consequence. However, although there isn't a material consequence, the time differential of the interaction plays a significant role, and that differential yields a consequence. Thus, a BTBB interaction is observable and measurable. In BTBB interaction we interact with a reality that, though pervasive in ATB reality, is distinct from the ATB universe.

With these caveats, then, let's begin our search. We'll start with a clear definition of the criteria that demonstrate interaction with BTBB reality. Then we'll search human experience to find those aspects that meet the criteria. As we perform our search, we'll tackle the issue described above. That is, if an experience meets the criteria, it doesn't necessarily mean that the experience is interaction with BTBB reality. Arguing that since all dandelions have yellow flowers, any yellow flower is a dandelion isn't a valid argument. And just as there are yellow flowers that aren't dandelions, there may be human experiences that meet our criteria that aren't BTBB interaction. Thus, we'll have to establish that the human experience we're considering has a unique character about it that can only be explained by identifying it as BTBB interaction. This will turn out to be a difficult challenge, but we'll give it a try. First, however, let's define our criteria.

There are two criteria that we've already assigned to interaction with BTBB reality: timelessness and nonmateriality. Timelessness comes from the drastically different speeds of the two realities, our everyday, ATB reality and the BTBB reality. Nonmateriality stems from the fact that Before The Big Bang our universe didn't exist. Matter as we know it came after the Big Bang. As matter accumulated, our universe began. At this point I'll offer a third criteria born of the pervasiveness referred to earlier—universality. Since all of our everyday reality was born from BTBB reality, that reality is fundamental to our ATB reality, and any interaction with it must apply universally. The interaction can't be a special case. As we gain a better understanding of timelessness and nonmateriality, we'll learn more about this third criteria. For now let's begin by defining the most difficult of the three: timelessness. This will take some effort, so please be patient. Timelessness provides the measurable criteria for BTBB interaction, and its thorough understanding will be critical in the identification of that interaction.

Timelessness

The explosion we label as the Big Bang was just that, an explosion. It was catastrophic, so catastrophic that energy became matter. The timeframe of BTBB reality (the rate that clocks move) must be vastly different than earth's, just as the timeframe of that speeding rocket ship off to another galaxy is different. That rocket ship, by the way, is going so fast that earthlings can't interact with it on a timely basis. By the time a message from earth reaches the ship, many generations on earth will have come and gone, because the rocket ship is traveling at almost the speed of light. Earth's radio waves catch up with the rocket ship long after earth has sent them. A logical conclusion is that the same lapse applies to BTBB reality. If there is a significant difference in the rates of time between two realities, practical interaction can't take place. Using this argument would force the belief that we ATB creatures can't interact with BTBB reality.

There is a difference, however, between earthlings interacting with a fast rocket ship and interacting with BTBB reality. The rocket ship is a part of our reality that's split off, heading for another galaxy at a fast speed. To compare interacting with it to interacting with BTBB reality, there would have to be omnipresence in both instances. But since the rocket ship and earth are separated in spacetime, that omnipresence can't occur. That isn't the case with BTBB reality, if BTBB reality is embedded in ATB reality. Such a circumstance allows for direct interaction with the two realities, since there are minimal spatial consequences in the interaction. If there could be pure BTBB interaction, there would be no spatial consequences whatsoever, but the interaction cannot be pure because of the ATB reality present.

Therefore, a key to understanding timelessness is to understand the difference between witnessing the time differential within a timeframe, such as the rocket ship example, and between two timeframes embedded in each other, such as would occur in any interaction we'd have with BTBB reality.

To better define the difference, let's step aboard that fast rocket ship that was heading to a far-off galaxy one more time and look around, checking for timelessness. First, we notice that the designers of the spaceship had enough classical thought to insist that the official clock next to the captain's chair be analog. We set our watch by it and wait a few minutes, noting that our watch runs exactly as it did on earth. The minute hand on our watch exactly matches the minute hand on the captain's clock. This doesn't change as the captain orders a speed increase that propels the ship even closer to the speed of light. After a small snack and casual conversation, we sit back and listen to the latest news from earth and hear two congressmen squabble about whether a new fuel cell system should be taxed at a higher rate so that funds for education can be increased. We turn our attention back to the course we've set for ourselves and note that we've made marvelous progress in the first month of our voyage. In fact, at our present speed, we'll reach our destination exactly when we'd hoped. We're transmitting information on the status of our mission back to earth, but we realize that it will now take almost a month of our time for the information to reach earth. Every day the delay becomes longer, but we knew that would be the case, since our ship is traveling so fast.

To understand the difference in timeframes, let's say the rocket ship has an emergency. The propulsion system begins to act up and, despite all the expertise and equipment on board, the captain concludes that the only way to save the passengers is to return to earth. The trip will take more than a month, but that is the only chance the ship has. The new course is set and the rocket ship heads back.

A month later the earth is in view. We know we're in for a shock, and the communication we've recently had with earth confirms that prediction. Nothing we knew, nobody we knew is there. In fact, our journey was recorded in ancient archives that apparently have been lost. Instead of a welcome, we're going to be treated as aliens from outer space. Our garb will be that of centuries-old dress, which in itself will be a curiosity. Our relatives will be dead, their children will be dead, and their grandchildren will be dead. What was a couple of months for us was several centuries for those left behind. This isn't a surprise. We knew our clocks would run slower. Now that we've returned, the time difference has been confirmed. It was there, even though neither those on earth nor the passengers on the rocket ship experienced the difference. It wasn't noticeable until we returned and the interaction took place. But when the rocket ship landed and the two realities met, the dramatic difference surfaced to such an extent that both realities, the earth's and the rocket ship's, appeared foreign and threatening to each other.

This isn't the case when considering interaction with BTBB reality. There's no delay in witnessing the time difference. For any interaction, BTBB reality must be embedded in us, which means there is that incredibly fast timeframe that suddenly surfaces, makes an amazing imprint, and then fades to the background as ATB reality's timeframe reappears. That is, time in BTBB reality is so much faster than it is in ATB reality that there's no real comparison in human perception. The BTBB interaction is unique.

Our lives are comparatively long, not short, when we use the pace of BTBB reality for the comparison. We have ample time to accomplish what we commit ourselves to do, provided we don't waste the time allotted to us. Perhaps more important, the pace of our time is slow enough that we have the opportunity to measure, analyze, and understand how our universe works. We're witnesses to all of earth's changes and our fellow creatures' adaptation to those changes. We may even be able to influence what comes next.

This ample, leisurely time isn't the case when we consider BTBB reality, however. If we could communicate with BTBB reality at this very moment, we'd see in that moment **its** equivalent of the period from the first man walking the earth, to a spaceship disintegrating on reentry, to our own reading of this entreaty attempting to describe the difference in the speeds of the timeframes. That is, we'd see all of that extended time period happening in no time at all.

The critical point is that the rocket ship example isn't the same for a situation where we're immersed in BTBB reality. For the latter case, BTBB reality encompasses us, which gives us the capability to have the communication the rocket ship couldn't. Thus, in interaction with BTBB reality, we would see virtually all of its reality at once.

That's a bold statement. Imagine being able to see all of BTBB's reality in a moment. Imagine its impact. We'd have an overview that would be so dramatic that the experience would certainly astound us. Contrast such an overview with the comparative narrowness with which humans view ATB reality, where we experience only the present slot of spacetime. We can remember the past, and we can contemplate the future, but the past is gone and the future hasn't come yet. BTBB reality is no different, of course, yet any interaction with it from an ATB viewpoint yields a much wider perspective. The past, present, and future of a BTBB perspective are all included in the ATB perspective.

This astounding observation must be tempered to some degree, however. The range of BTBB reality that can be observed is limited by the imperfection of the BTBB interaction. That is, the only way to totally experience the wide overview that the difference in rates of time dictates would be by free and complete interaction between the two realities. We would not only have to find that embedded BTBB reality, but we'd have to somehow interact with it as purely as we interact with our everyday, ATB reality. Unfortunately the physical aspect of ATB reality means that there must be space within any reality we experience. That space, in whatever shape it has, is a barrier that disallows perfect communication between ATB reality and BTBB reality. Thus, our quest for evidence of

BTBB interaction must be limited to an imperfect interaction that allows for ATB space. Our viewpoint will be much wider when considering BTBB reality than it is when considering ATB reality, but the width of the viewpoint is dependant on the quality of the interaction.

So the timelessness of BTBB interaction means experiencing a wider scope of reality than we normally do when interacting with ATB reality. Timelessness also means experiencing a different rhythm. To understand the rhythmic aspect, let's go back to the rocket ship analogy.

A day of our life on the ship is quite close to a normal day on earth. We spend about a third of it asleep, eat three meals a day, and spend some time at leisure and exercise to keep our muscles in shape. The rhythm we had on earth continues. We plan to pass on this rhythm to our offspring even though those offspring will never set foot on earth. Why wouldn't we? This rhythm is part of our heritage, and it's worked well to keep us healthy and our minds sharp and productive. We want our children to have the same advantages that we had, so we'll train them accordingly.

Earth's cycle is built-in and fundamental to the passengers on the rocket ship. Similarly, the BTBB timeframe should be fundamental to our universe. It was in existence before our universe came into existence. Our universe was born from it. Thus, the BTBB timeframe is ingrained somewhere in this existence, just as earth's timeframe is ingrained in the passengers of the rocket ship. Even if the earth was destroyed and nothing was left, the passengers would still have the legacy. So too with BTBB reality. We don't know exactly what happened to the BTBB reality other than that we're the result, but that fact doesn't discount the possibility of our experiencing the BTBB timeframe. The BTBB timeframe may be more subtle. It certainly is more remote. More important, however, is its fundamentality. The timelessness that we are considering is the rhythm of BTBB reality within our ATB timeframe.

So in timelessness we have a wider view of a reality and a different rhythm. But that isn't enough. Most readers will still have difficulty identifying timelessness at this juncture. Perhaps there are even those who will claim that making such an identification is beyond human capability. That opinion certainly isn't unique. Some believe that humans are incapable of fully understanding the universe. The ones who make that claim question human intellect and limit mankind's understanding to only those aspects that people can experience.

Yet Einstein hadn't experienced the relativity of time before he proposed it. Many scientific theories have contradicted our everyday experience only to be proven correct later. The pessimists are wrong. Humans have proven over and over that mankind has the ability to escape the walls that everyday reality presents. Just as Einstein could conceive the spacetime continuum, so too will others conceive new theories on how the universe works. Mankind has used and always will use imagination and creativity to stretch beyond the perceptions of everyday experience. Humans in the future will, as they have in the past, make discoveries that transcend parochial limitations. So although pure interaction with BTBB reality may be impractical, we can search for hints of that interaction and use timelessness as a tool to reach our target as closely as possible, all the while functioning in ATB reality.

Great ideas are simple, not complicated. For example, rules governing the mathematics of Newton's Calculus come from the simple idea that continuous functions—those mathematical relationships that have no gaps—can be calculated by considering them in infinitesimal increments. Adding those increments up theoretically, rather than actually doing the arithmetic, is not only possible but far more efficient. There is great simplicity in that. Those who have difficulty with calculus do so because of a poor background in basic mathematical skills and not because of the complexity of calculus. The idea that our time is much slower than time in the BTBB timeframe is a simple idea. There's nothing original in it, of course. It's merely a logical extension of Einstein's Special Theory of Relativity.

We're in a timeframe that's impossibly slow compared to the BTBB timeframe which is the logical consequence of being the result of the Big Bang. Our slower timeframe gives us a different view of ourselves as compared to a BTBB view, but it doesn't prevent us from imagining how the BTBB viewpoint is different from our own. The differences in the points of view can lead to a better understanding of our whole reality.

The definition of timelessness is even more difficult to accept than the relativity of time, and the only confirmation that it exists is that very relativity. So if you're confused at this point, don't despair. We'll further explore the definition of timelessness until we've established enough confidence to recognize it when we find the interaction with BTBB reality that we're looking for. Next we'll look at the implications of confronting a much faster timeframe.

Chapter 5
A Deeper Look at Timelessness

What does it mean to experience the rhythm of a faster timeframe? How would a human react to seeing time literally fly by? In order to answer that question, let's explore differences in rates of time and see the role BTBB reality could play.

Remember, for BTBB interaction we're like the passengers on the rocket ship whose time is slower than earth's. For that analogy, the rocket ship is ATB reality and earth is BTBB reality. If there could be instant communication and someone in the BTBB reality could see us, the time difference would be dramatic. Generations would pass in the BTBB reality while we chewed on a mouthful of food. On the other hand, if we, the passengers, could look back on earth, everything would be happening so fast that it would be a blur, almost as if everything was happening at once.

Understanding time means understanding change, but if the rate of change is impossibly fast, it distorts our view of change. The film industry uses that phenomenon, of course, since movie films are nothing more than a series of still photographs taken in rapid succession which are played back so fast that they appear continuous to an audience. TV uses the same technique, except that it replaces pictures with fast scanning electrons. In both cases the super-fast changes become a blur and mask the reality. The result, however, is that the alternative reality appears continuous, and we enjoy the entertainment.

Changing information faster and faster may enhance the quality of a movie and provide better entertainment value, but that in itself doesn't make a substantial contribution to our definition of timelessness. Fast change merely fools our senses. So if time is that much faster when interacting with BTBB reality, what is the best way to portray this timelessness to get a sense of it?

Perhaps an analogy would work, so I'll use one that has the best chance of gaining an audience. I'll use sex. Any author who chooses physics to convince readers of anything must fight to get attention, so a savvy editor would probably insist that sex be included somewhere. How else would the thing sell? Certainly sex will attract a larger audience than, say, the curvature of space. So let's look at interaction with BTBB reality and see how sex might look in that interaction.

Before we begin, however, one aspect of the time difference between the two realities must be repeated. Within the reality, whether it be in a fast rocket ship, on earth, in the BTBB reality, or in any other reality, time appears to be the same. Clocks move at a different pace, but to an observer within a timeframe the clocks move normally. The reason is that everything in a timeframe moves at the timeframe's rate, clocks, metabolism—everything. Time may be different in comparison to another timeframe, but it moves at the same rate for a person on a fast rocket ship as it does for a person on earth. It's only when there's interaction between the two timeframes that the difference surfaces. Thus, if there were sex BTBB, it would be the same for the participants as it is for us. On the other hand, if we look at BTBB sex from our timeframe, we would be like the passengers on that fast rocket ship who—if they could look back on earth—would watch generations live and die in moments. We'd have that omnipresence. Of course it's impossible to be such a witness, since the rocket ship is going so fast. But we're talking about sex, here. Fictional sex. Sex from an omniscient point of view. So relax and allow yourself to consider BTBB sex from our ATB viewpoint.

First, foreplay is so quick that it isn't even noticeable. (Now you know why there's a shortage of female physicists!) There's no lengthy meeting of the eyes, quickening of the pulse or tingling of anticipation, at least none that's measurable. The coupling appears immediate with all the intensity, passion, and lust of climax. Yet the coupling is just as quick as the foreplay. And, like everything else, there is no lasting, post-climax bliss, no enduring delight in memory or lingering, exhausted joy. There are all of these but none of the duration. Duration isn't an issue since, like the climax, everything is instantaneous and fleeting. But everything **is** there with all its intensity and enjoyment; only it's all there immediately and just for a moment. There's no buildup, no letdown, no nothing except the intense beauty of the experience.

The above description of BTBB sex from the ATB viewpoint isn't very appealing, is it? We are used to sequence and an occasional pause in between steps for contemplation and satisfaction. To imagine even the most pleasurable of physical acts outside of time is repulsive—even for men. Our whole being is embedded with time, and that is the way it should be. Our physical self is matter, and matter is constantly changing. We measure with that change, enjoy with that change, contemplate with that change. In short, change is what life is. The change must be sequential, however, and have a reasonable rhythm. If it doesn't, we regard it as chaos. So if the change is so rapid in interaction with BTBB reality, would we find the experience so outrageous that we wouldn't appreciate the interaction? Or would it be like an engrossing movie where we don't notice the still pictures changing, only the result? That is, is interaction with BTBB reality similar to witnessing chaos, or do humans actually only perceive and use the result? Or, as a third alternative, do humans understand each component of the "blur"? How exactly do human beings experience timelessness?

From our everyday-reality viewpoint, an hour is a normal hour, but, from the BTBB reality viewpoint, that hour is so long that it's interminable. On the other hand, from the ATB timeframe viewpoint, clocks in the BTBB timeframe would appear to spin in that "blur." This fast-moving time, when so much can happen so fast, is the timelessness that we're looking for. An interval must pass so quickly that it's virtually immeasurable, and, for time to pass that quickly, there has to be an immense change.

To understand how humans experience timelessness, let's look at the just-mentioned condition a little more closely. Time is change, but it's also duration. If absolutely nothing changes for an hour, the hour has still passed. Of course, in that hour some things **did** change. The sun moved. Humans and other living things aged. There's no such thing in our experience where absolutely nothing changes in a given amount of time. Duration allows the change to occur at a given rate. To identify interaction with BTBB reality, we must witness something that happens in that duration at a much faster rate than it should by our normal standards. The time (duration) it takes in our reality has to be almost zero, yet the change must be so dramatic that we might have expected it to take generations to occur.

Understanding timelessness means understanding that BTBB reality is embedded in ATB reality, thus minimizing the spatial considerations for comparing the two timeframes. That means witnessing a phenomenal amount of change in a minimal amount of time. The idea that timelessness means immense change in virtually an instant is difficult to comprehend. Nevertheless, that is what timelessness means for humans experiencing BTBB interaction. It means experiencing a wide view of a reality that has a different rhythm, and that rhythm is so fast that it's difficult to grasp.

Fortunately, defining the criteria of nonmateriality and universality is much easier than defining timelessness, but, before we move on to those criteria, we must clear up the issue of the human capability to experience timelessness. Exactly how do humans perceive interaction with BTBB reality? In the next chapter, we'll define a reasonable way for humans to experience and cope with timelessness.

Chapter 6
Timelessness in Interaction

Each slot in our everyday spacetime is unique, and once we occupy a slot of spacetime, we will occupy that slot in the universe forever. It is that aspect of our everyday reality that provides the essential building blocks for the timelessness of interaction with BTBB reality. Once we understand those building blocks, perhaps we can tame the chaotic nature of timelessness. First, though, the concept of forever occupying a slot of spacetime may be confusing, so let's do a little review to clarify that idea.

The reason our slot of spacetime occupies the universe forever is that there is the possibility of another viewpoint in a far-off galaxy witnessing that slot sometime in the future, just as we witness events of far-away galaxies years after they've actually happened. Our NOW might be many light-years away from other witnesses on other planets in other solar systems of other galaxies. To these witnesses, the earth might still be inhabited by dinosaurs and wouldn't be worth the visit. (Obviously, these fuddy-duddies wouldn't have the imagination of a human ten-year-old.) These far-away observers wouldn't know that we've been born yet, because our life-style wouldn't be observable. There surely wouldn't be any evidence of an author trying to sell the outlandish notion of interaction with BTBB reality. Our contemporary existence wouldn't appear until long after we were dead and buried.

Yet eventually, if they looked in the right direction, those far-away witnesses would see human civilization rise. They or their descendents would witness the rise and fall of the Roman Empire and the European discovery of the Western Hemisphere. They'd watch as that discovery led to skyscrapers, space travel, and smog. Other witnesses in other galaxies even further away would witness the events later. Still other witnesses, who might be closer, would witness the events after they occurred but sooner than the first set of witnesses. The point is that our life on earth is a series of spacetime slots that, for us, are sequential and isolated to our planet. They are our NOWs. But these NOWs will appear later on to other witnesses. How much later depends on how far away the witnesses are.

We don't include this endless aspect of spacetime slots in our NOWs. We live in our own little niche of spacetime that we label as life. We base our time on a 24-hour day, the time it takes our planet to make a complete rotation. We note the activity of far-away galaxies, combine what we witness with our NOWs, and proceed. But our observation of that activity is delayed, and we are, therefore, oblivious to the actual reality that exists in those galaxies.

For argument's sake, however, let's say that witnesses from a distant galaxy invade our spacetime. If the galaxy was far enough away, those witnesses would have traveled at a high rate of speed to get here in a reasonable amount of time. Their speed means they would have had a slower time, and that different timeframe would be evident when the witnesses arrived. The perspective of these witnesses might be extraordinary. If the visitors had glanced at earth while on their journey, they would have noticed the changes as they approached. In totality the visitors would have seen the events of our everyday reality occur at a far different rate than we had.

How would we interact with the visitors? If we could communicate, we'd exchange views about how the rate of change differed. The rate of change on earth would have been far faster for the space travelers than it had been for us. The visitors might have left their home just after the last ice age and arrived when we'd completed an exploration of Mars. Imagine how impressive our progress would be to the visitors, and imagine our interest in being able to discuss an earth enveloped by glaciers.

Remember, our own time slots, the ones of our lives, don't end. The information is out there and remains out there even after our death, just like the information showing glaciers covering the earth. The slots travel through space and are available to any receiver powerful enough to pick them up. Just because we haven't interacted with some witness from another galaxy doesn't mean that a witness isn't out there. That witness could watch my existence years after I'm dead. My slots of spacetime linger. They stay out there, as Brian Greene points out, forever.

So what does that mean? Unless somehow we could hop around from one spacetime slot to another—the old time machine issue—those slots have no application in our existence. For us, what counts is occupying the spacetime slot that we occupy and no more. However, say we could hop around, and say we could do so with almost infinite speed. If so, we could interact with all of these slots at once. That would be interacting with ATB reality in a timeless fashion. Similarly, **when we speak of timelessness when interacting with BTBB reality, it's as if we could capture essentially all of the images, that is, all of the time slots of BTBB reality, at once. That's the timelessness we would experience!**

Again, such timelessness is mind-boggling. It represents a perspective that seems foreign to human experience. Trying to evaluate timelessness as defined here presents a challenge to the human intellect. Yet experiencing such timelessness may be possible. Although it would take time to absorb all those BTBB images at a reasonable pace, a human being might be able to accept the result, the final "picture," and work with it. That is, humans could react to and use timelessness just as they accept the rapid display of pictures when viewing a movie. When they do so, they ignore the inundation and witness the result. Thus, interaction with BTBB reality would be like viewing a movie where the chaotic mixture of a huge number of pictures is replaced with an orderly view of what appears to be a "moving" picture. Our minds would absorb the "movie" and ignore the "stills." In this way it would be possible for humans to experience virtually all of BTBB reality at once without being bombarded with details.

Using this approach allows a reasonable method for humans to experience BTBB interaction. That is, this approach means that we don't experience a barrage of BTBB timeslots, but a panorama of timeslots that is neither haphazard nor chaotic.

Experiencing timelessness means experiencing a unique moment unlike anything ATB reality presents. It's experiencing a great amount of change—all those images of BTBB reality—in a very short time. Humans can have that experience, but in order for the event to be meaningful, the images must be clear, so they can be absorbed distinctly. Viewing the experience as a panorama and not a blur provides that order.

Even though timelessness is difficult to understand and accept, it presents an interesting twist to the human viewpoint of our universe. When a witness in the ATB timeframe sees change in the BTBB timeframe, the change is so fast that it's immeasurable. This means that the eternity of the BTBB timeframe is an instant in our timeframe. Eternity is an instant!

Let's apply this idea to an old argument. Creationists and evolutionists have been debating with each other ever since Darwin first presented his theory. Creationists maintain passionately that God created the universe and everything that's in it. To imply otherwise and claim that nature stumbled along on its own is total nonsense to a creationist. On the other hand, scientists maintain that species have evolved over time, changing due to the requirements of their environment, not as a result of divine supervision or intervention. Members of species that didn't have the necessary mutation didn't survive. Members that had the mutation did.

The argument between these views rages to this day, and it becomes even more passionate when the debate concerns what we should teach our children. Evolutionists claim that creationists want to teach religion in schools, while creationists say that godless evolutionists want to inflict their cancer on our youth and deprive them of knowing about the divine care that blesses our existence. Evolutionists scoff at the accusation. To them, ignoring evolution is ignoring scientific discovery. Creationists in turn scoff at such scientific certainty, pointing out that history is full of incorrect science, and besides, evolution doesn't fully explain earth's creatures. When it comes to science, they proclaim, nothing is infallible. Thus, teaching evolution as a fact to children is wrong until it can be proven that it explains all of life. On and on the argument goes, neither side willing to yield, and both sides absolutely sure of themselves.

Perhaps both sides are correct. Evolutionists are considering creation in earth's timeframe, since that is the only reasonable timeframe for legitimate science. They are required to measure and verify facts, which means that they have no other choice. However, in making that choice they ignore another possibility, the timeframe that would exist if evolution is considered from a viewpoint that is equivalent to our view of BTBB reality. This would be the viewpoint of a fast rocket ship or of any entity with a significant velocity compared to ours. In that case, this other viewpoint would have a slower timeframe where billions of years in earth's time would only be a faction of a second in the new timeframe.

If the development of birds growing wings or carnivorous beasts standing on two legs took only moments, evolutionists might draw their conclusions differently. Perhaps they would think they were witnessing the steps of creation rather than the plodding procedure called evolution. Perhaps not, but considering interaction with ATB reality from a viewpoint equivalent to our interaction with BTBB reality provides meaningful insights that could help our understanding of life. The controversy between creationists and evolutionists is a good example.

Of course, comparing views based on different timeframes is controversial, and any result will not necessarily be accepted. In the example of evolution, for instance, we can imagine how debatable any conclusion would be. Evolutionists would challenge a proposal to use some other timeframe for making a judgment, considering it as a distraction inapplicable in their work. Creationists could be equally concerned, arguing that introducing another timeframe would only confuse the issue and simply be another way to take a Supreme Being out of consideration. Looking at timelessness may explain why evolution is slow for us but fast from some other viewpoint, but for adversaries it probably wouldn't squelch the argument unless they both agreed to open their minds a bit.

Face it, thinking about viewpoints from other timeframes is challenging. Considering everything happening essentially all at once isn't satisfying. It would be like evolution taking place in the wink of an eye. Our lifetime would be so short that it would be trivial. But life isn't trivial to us. That's why we need duration. We function at a pace that allows for reason, judgment, and planning. The scientific method is based on the pace of our everyday timeframe.

For interaction with BTBB reality, however, the timeframe is completely new and foreign. Thus, timelessness is a difficult property to comprehend and will be difficult to recognize. Our everyday consideration of time is real and works as far as day-to-day life is concerned. But now we know that time is relative, so looking at reality from other timeframes can be just as real. Of course, once time is taken out of the equation, the whole world is completely different. We do have one advantage—we can use or minds and our imagination. We can look for that moment when a huge gulp of time is condensed into a moment.

There are many, many timeframes. In fact, once the possibility of high speed travel is accepted, the number of timeframes is endless. However, any timeframe other than the one we normally use is impractical for everyday considerations. Further, in ATB reality we can only slow time down, not speed it up. That is, we're only able to accelerate not decelerate. If we tried to decelerate, we'd only accelerate in another direction. We can't go slower than we're already going, so we can't speed time up. We can only accelerate and slow time down. Other timeframes may provide a solution for traveling to far-away galaxies, but they do nothing to explain our NOW in the universe. They don't provide the amazing insights that an interaction with BTBB reality can provide, since they don't allow for viewing a vast number of timeslots in a short amount of time.

Our everyday reality is the result of the Big Bang. Our time must use the BTBB timeframe as its basis. **We accelerated from that point, so our time is slower than that timeframe. It's the BTBB timeframe that permits interaction with a large chunk of reality in an instant. That's why the BTBB timeframe is so special and unique.** Nevertheless, understanding and working with this timeframe is incredibly difficult. We must keep reminding ourselves that, in comparison to BTBB reality, we live in a very slow timeframe. Yet there was a timeframe BTBB, and that timeframe still exists somewhere. The BTBB timeframe is much, much faster than ours and this faster clock, by definition, must surface in any BTBB interaction.

We live in a timeframe that we understand, even though it took mankind a long time to define a day as 24 hours and then measure it accurately. Since it took so long to understand the timeframe of our existence, think how difficult it will be to understand an interaction with some other timeframe, especially since that other timeframe is so foreign to our normal experience. In everyday reality we're used to progression and duration, and if one of those traits is violated we feel cheated. Take the four seasons of the yearly cycle as an example. We look forward to the heartening feeling that spring wildflowers bring and revel in the joy of springtime warmth. We listen to birds sing and watch them show off their young. We witness the change as that warmth grows into the heat of summer. After months of hot, sweaty days, we see the green draining from leaves, leaving the vivid colors of fall. The clean, crisp air hints of the cold that is to come. The leaves fall, but we bask in the confidence that they'll be back. Before they return, though, we look forward to the first snowfall that brings a hush and a whiteness to our lives.

So if there were a year when the leaves appeared, turned from green to orange, and fell to the ground, all at essentially the same time, the year wouldn't make sense to us. If birds built nests, gave birth, and fed their young with no time in between, we'd think it nonsensical as well. It's almost impossible to think of the incredibly fast change that interaction with BTBB reality implies. Just contemplating it is overwhelming, confusing, and frustrating.

Where's the progression? Where's the duration? Logic tells us that cause has its effect, so if the cause and effect are essentially simultaneous, there is something wrong with the reality. It's an incomprehensible reality that is beyond human understanding.

Yet if we interacted with BTBB reality, we would experience a similar sort of timelessness. From our ATB point of view—all that happening in so short a time—the experience would be startling. That huge amount of change, however, is the key to identifying BTBB interaction. It must be an experience where our normal pace of time is violated, replaced with the new rate of BTBB reality. And we must use the result and accept it as a normal human function. We may be surprised or astonished, but we won't have experienced nonsensical chaos. Remember, BTBB reality is the fundamental reality from which our reality began. The BTBB reality is our foundation. That's why the apparently "startling" experience will appear quite natural.

Let's review this one more time. The ATB building blocks of slots in spacetime indicate change, and change defines time. A measurement of time is a measurement of change. Just as we define a day as the time it takes for the earth to make one complete rotation, we define everything else with respect to that duration. If the earth didn't rotate, we'd use something else as the basis for our rhythm, since we need a standard for duration. We'd use that standard to allot time for sleeping, working, and playing, and we'd base our day on that standard. One thing is certain, whether the standard used was the movement of a heavenly body, sand passing through a narrow opening, or the rate of decay of a radioactive material, that standard would have duration. There can be no such thing as time without duration. If time stops, there is no movement of the sun, we don't age, the temperature is constant and everything else is frozen in one of those spacetime slots.

The timelessness of interaction with BTBB reality, however, means the opposite of time stopping. It is tremendous change in minimum duration, change so rapid that there appears to be no distinction between the slots of spacetime. Such interaction with BTBB reality seems to violate our sense of time, duration, and reality. Yet if we have that interaction, it must appear normal to us, or we'd be asking ourselves what was happening. The reason it doesn't appear odd is that humans have learned to work with BTBB reality, which is the fundamental reality of our universe.

Einstein's Special Theory of Relativity appears to violate our human experience, but, as we understand the concept and grow with it, it becomes more "natural." Identifying interaction with BTBB reality is similarly challenging. It's subtle, so subtle that we normally don't recognize it when it occurs, even though it's as different as it is. This feature makes using timelessness as a criteria for BTBB interaction so intimidating. Nevertheless, it is the one measurable feature of that interaction and, therefore, very precious.

Thus we have defined the most difficult component of interaction with BTBB reality—timelessness. It means immense change in minimal time. It means that in an interaction our instant is BTBB reality's eternity. It means the equivalent of experiencing all of mankind's time slots virtually at once. And it means humans have the experience in an orderly, natural manner that may be astonishing but won't appear magical or mystical. With that difficult concept in mind, let's turn our attention to nonmateriality and universality.

Chapter 7
Defining the Second and Third Criteria—Nonmateriality and Universality

Nonmateriality

Our universe is made up of matter and energy. One of Einstein's brilliant discoveries was that these components are merely different aspects of the same reality. His famous equation, $E = mc^2$, defines that relationship. Notice that the square of the velocity of light is the factor that connects the two. Why the velocity of light instead of some other factor? The answer is that the total energy of a system includes both its mass and its velocity. Velocity is a directed distance measured over time, and we already know that time is dependent on velocity relative to the velocity of light. The fraction v^2/c^2 comes into play. Thus c, the velocity of light, is critical in defining the relationship of mass to energy.

Defining nonmateriality is straightforward. There's nothing mysterious or mystical implied. All the theorizing about spirituality mankind has done over the years may be interesting, but it has no use here. Nonmateriality is merely the recognition that the BTBB universe was essentially energy, and that some of the energy evolved into matter as the result of the Big Bang. Nonmaterial means essentially pure energy. Thus, interaction with BTBB reality must, to a large degree, include energy and exclude matter.

In our search for interaction with BTBB reality we will insist that the interaction be somehow measurable. We will exclude the mystical. Yet if nonmateriality is without matter, how can we measure its physical energy? That is, what can we use for a reference if there is nothing material to measure? Clearly, it's a formidable challenge where time plays the dominant role. Fortunately nonmateriality, despite that challenge, is the easiest of the three criteria to grasp. It means our reality as energy, a similar state to that which existed BTBB.

Universality

The third component of BTBB interaction is universality. The requirement that BTBB interaction must have universality isn't immediately obvious from comparing the two realities, our everyday reality and the BTBB reality, and, therefore, it requires some effort to define. To understand where this component comes from, let's return to our rocket-ship analogy and look at it from the earthlings' point of view.

The passengers who flew off into space came from earth. We knew them before they took off as human beings just like ourselves. They could have been relatives, friends, or merely known adventurers, but no matter how well we knew them, we understood their humanity. We could anticipate their reactions and thoughts, and we had a certain commonality with them in approaching problems and planning for the future. We can't divorce ourselves from them just because they chose to travel at a high velocity to reach another galaxy and, thus, in a sense, chose a different reality from our own. Their time might be slower than ours, far slower, but there's still a connection between our reality and the rocket ship.

And, just as we knew them, they knew us. They might speculate about the time difference, since they'd be well-versed on Einstein's theory, but they'd observe their clocks as acting the same as before they took off. As discussed in Chapter 5, they'd have the same outlook as they had had on earth. To them, their new rate of change, their time, would be the same as it had been on earth. They'd know it was much slower than earth's time, but it wouldn't feel that way. They'd carry earth's heritage along as well, just as we would carry their legacy on earth. The two realities would be coupled in a sense, since both came from the same source.

Again, remember that since the rocket ship is going so fast, there can't be any meaningful interaction—communication—between it and the earth. If timely communication could occur, the dramatic differences in the rate of time would be evident. But timely communication **isn't** possible. The same issue applies to interaction with BTBB reality, so we've concluded that if we find the interaction it will somehow be a special case. The question is why could there even be a special case? When considering the possibility of interacting with a reality with a significant difference in velocity, what's the difference between earthbound interaction with BTBB reality and a rocket ship's interaction with earth? In chapter six we gave the answer to that question: it's because BTBB reality is embedded in ATB reality.

That's where universality comes in. The Big Bang affected our whole reality, not just a segment of it as a rocket-ship mission would. Our universe came from the Big Bang—all of it. Thus, we carry the BTBB heritage just as the passengers of the rocket ship carry our heritage. The difference is that our BTBB heritage is basic to our own reality. BTBB reality gave us our present reality, and, in that sense, heritage is too weak a term. BTBB reality gave us our complete reality. Thus, if we find interaction with BTBB reality, it must be universal. That is, it must be all-inclusive and embedded in ATB reality. It can't be a special case or an abnormality requiring unique insight that isn't shared throughout humanity. All humans must have the capability to have the interaction, although the intensity may vary depending upon individual talent.

* * *

This ends our discussion of the criteria we expect to find when we interact with BTBB reality. We've defined three critical components: timelessness—vast changes in minimal time; nonmateriality—pure energy; and universality—a commonality existing within ATB reality. This commonality is pervasive and inclusive. Now we can begin our search, keeping in mind the caveats we noted in Chapter 4. First, there won't be a "material" BTBB reality that yields a physical consequence. Second, the interaction will be imperfect. Third, there won't be a specific "place" where BTBB reality resides. And finally, fourth, the interaction can only be completely explained taking BTBB interaction into consideration.

Whatever the result of our search is, there's one certainty— it will be challenged. The challenge could come from scientists who consider the exploration as a threat to their legitimate work or a corruption of their valid interpretations based on the ATB universe. Or challenges might be theoretical, from philosophers who prefer to concentrate on ideas that are independent of scientific and mathematical complications. All challengers would probably protest that using a basic observation such as the relativity of time to solve anything as complex as the human condition is preposterous. We welcome those challenges. Questioning proposed ideas can suggest adjustments and contribute to mankind's progress. So challenge away, but please, attempt to understand the possibility of BTBB interaction first.

To find interaction with BTBB reality that meets all of our criteria in a pure state is not just a formidable challenge, but, as we have already noted, an unreasonable expectation. All of the interactions we discover will have a limitation of some sort in addition to the other conditions. This fact shouldn't be discouraging, however. A dogged pursuit for as many examples of limited interactions as we can find has the potential for greatly improving our understanding of the total reality that envelops our existence. And, if we adhere to the rule that the interaction must be observable and measurable, we'll eliminate the subjectivity that often surfaces in any pursuit of reality beyond the normal, everyday, ATB reality .

Of course, if BTBB reality is embedded in our existence, we should see evidence of the reality everywhere. It may be in ourselves, but it also should be observable in the physical reality outside of ourselves. So before we look within, let's look at the external reality that humanity has already accepted and see if it displays traces of the criteria we're looking for. The search will be tricky, since physical reality is material and what we're looking for is nonmaterial. An additional issue will quickly surface as well. Remember that unless the reality is embedded within ourselves, there is that spatial consideration similar to the rocket-ship communication problem. Space limits observation, so the principal measurable component, timelessness, is unlikely to surface. Nevertheless, if BTBB reality is universal, the reality outside of ourselves should provide some hint that it's there. Accordingly, we will spend the next few chapters looking for BTBB reality outside of ourselves. For readers who are people-orientated, we'll get to our main topic—human consciousness—after that. When we do so, however, we'll have even more confidence in our results, since we'll have some experience with embedded BTBB reality that we can observe outside of ourselves. So let's take a close look at our physical environment and see if any of it exhibits the criteria that we've defined. That environment, of course, is vast, and the quest raises a question—where to begin?

It's not as hard to find a starting point as you might imagine. You see, physicists have made amazing breakthroughs, and some of those breakthroughs actually appear, in one way or another, to meet the criteria we've defined. We'll be delving deeper into physics in the next chapter, so be warned. The crash course might be a lot to swallow, but we'll ignore the math and stick to the results that physicists have discovered.

Chapter 8
Similarities between Particle Physics and BTBB Reality

Physics covers a vast array of topics, but there is one discipline—the study of particle physics—that has identified properties which suggest meeting the criteria we've established for BTBB interaction. This observation shouldn't be a surprise. In our everyday life, matter and energy are clearly defined and separate, but that isn't the case in particle physics. There, energy and matter have an interplay that appears to have an effect on their characteristics. Something unique about reality occurs at the sub-atomic level—the laws that govern the macro-world break down. Exactly why, no one is sure, which is the reason physicists are currently working feverishly to find a theory that works for both the sub-atomic level and the macro-universe.

The Big Bang itself was particle physics personified. BTBB reality was sterile. Essentially, matter as we know it didn't exist, and, since there wasn't any matter to change, our timeframe didn't exist either. At the moment of the Big Bang, however, the universe, our reality, suddenly began. The change was as immense as it was sudden, the universe exploding at an incredible rate. And, as we have seen, the velocity of the matter created at that moment defined a timeframe much slower than that of BTBB. Our material universe along with our timeframe, the ATB timeframe, began. The much slower clock that we currently use to make our measurements began to tick.

The matter created eventually coalesced into stars and planets. Earth came into being, and matter morphed into life-sustaining substances that eventually allowed mankind to walk the planet. It took ages in our timeframe, but eventually mankind began to study our universe. Thus, by the twentieth century, humans were probing into the subatomic world, the same world that made up the universe at the time of the Big Bang. At that point scientists made the startling discovery that the laws governing macro-reality don't work in the micro-world. This discovery forced them to invent a whole new system to explain the subatomic universe. They invented quantum mechanics.

Some scientific discoveries are so complicated that, although they are perfectly valid, are difficult to comprehend. Quantum mechanics is one such example, since it uses challenging mathematics that inhibits the understanding of its amazing discoveries and predictions. There is no simple equation or argument that makes the ideas of quantum mechanics intuitively acceptable.

Such problems with scientific discoveries aren't new, of course. Classical science offers similar difficulties, so it would be a mistake to dismiss quantum mechanics simply because of its complexity. For instance, the mathematical concept of entropy can, for some, be almost as challenging as quantum mechanics. Although entropy was developed using macro-world theory, it appears to apply universally to both the macro and micro world, so that's another reason to bring it up here.

Entropy is a measurement of randomness, which seems simple enough. Like quantum mechanics, entropy uses probability to define its aspect of reality. The reason its mathematics is difficult to grasp for some undergraduate science students may be that very probability, which requires a certain talent to master. Expertise in calculus and differential equations doesn't necessarily mean an intuitive ability to understand variations of statistical theory.

The law of thermodynamics using entropy is quite innocent. It simply states that the total entropy of the universe always increases. Thus, the law is just another way of saying randomness must increase. That's nothing new, though, is it? Ice cream melts; it doesn't harden unless it's put in a freezer. If it goes in a freezer, a motor runs to keep the freezer cold. That motor generates heat, which means that molecules are moving more rapidly, that is, in a more random manner, and, since the freezer isn't 100% efficient, it generates more heat than it removes from the ice cream. The net result is an increase in randomness. The same idea applies to other situations. For example, if a glass falls and breaks, it doesn't mend itself. The less orderly broken glass is in a more random state than glass that isn't shattered. A law of thermodynamics states that if a sequence of events doesn't result in more disorder in totality, it can't happen. That is, molecules of the whole system—of the universe—will subsequently vibrate in a more and more random fashion for any process that's feasible to occur.

Although the mathematics of entropy isn't necessarily easy to grasp, the concept of randomness is reasonable enough. So if the law says that randomness always increases, and we compare that requirement to our personal experiences, we can accept the conclusion rather easily. Ice cream melts; it's an experience we first had as a toddler, and one we're not about to challenge. Entropy corresponds to everyday experience. Thus, at least with entropy, the results that the difficult mathematics support are intuitively verifiable.

Not so with quantum mechanics. Like entropy, quantum mechanics insists on randomness in the micro-universe. Rather than getting a definite answer to a problem, however, (one can calculate the increase in entropy) a quantum-mechanic solution merely gives a probability. So while entropy is a study of randomness, quantum theory is a study of the chances for a particular event within that randomness. That is, quantum mechanics gives the **probability** of a defined solution rather than a solution itself.

There are two mathematical structures that portray quantum theory, one using matrix theory and the other using an equation called Schrödinger's Wave Equation. This is heavy theory and no casual presentation of quantum mechanics could include its mathematics. So when we learn what scientists have found about the micro-world, we won't have the satisfying feeling that we had after we considered Einstein's Special Theory of Relativity. The ideas of quantum mechanics, however, are supportive but not essential to the subject at hand, so a condensed discussion should be sufficient.

Like Einstein, the scientists who advocated quantum theory had to work hard to sell their ideas. A huge debate blossomed over the authenticity of quantum mechanics, the loudest challenger turning out to be none other than Albert Einstein himself. After all, Max Planck, Niels Bohr, et al had developed a theory that led to astounding conclusions. Quantum mechanics predicted such remarkable and unfamiliar results that Einstein was convinced the theory had to be erroneous.

The subatomic world offers a multitude of surprises. For example, we're used to defining, measuring, and locating anything physical, the only restriction being the capability of the instruments doing the identification. Yet Heisenberg's Uncertainty Principle states that both the exact position (or state, such as spin) of a particle and the exact moment it's in that position cannot be known simultaneously. We can measure where a particle is, or we can measure when it's there, but we cannot make both measurements precisely and concurrently. Again, that goes against our everyday experience. Whether it's a building or a speck of dust, if we have the proper equipment, we can measure precisely where it is and when it's there.

Another strange property of the subatomic world is that the very fact that an observation has been made can influence the next measurement. That is, a particle can change characteristics depending upon whether a measurement is made or not, even if the measuring process calls for no physical contact with the particle. This result is unlike interaction in the macro-world where measurements affect the object under observation only when there is physical contact. This difference implies that once we interact with a particle, once our timeframes lock, the subsequent state of that particle is no longer the same, and that fact applies even if the observation or interaction is indirect.

The results of quantum mechanics' mathematics offer additional oddities and can stretch believability to the extreme. For instance, the math implies that if a scientist finds a particular electron in a laboratory, there is a finite probability that the next time that electron is identified, it could be anywhere in the universe—on the moon, in a different galaxy, anywhere. The highest probability, of course, is that it's still in the room, but it's not a 100% probability. The idea that it could be anywhere else in the universe makes the concept almost incomprehensible.

These remarkable and unbelievable results weren't the only problem for Einstein, however. After all, he had developed his own theories that challenged common understanding. To him the most damning aspect of the theory was that it did not fit in with his theories of the macro-universe. (In actuality, the two theories were developed essentially concurrently, but the controversy came when both ideas were developed far enough to sustain their arguments.)

The fundamental phenomenon that quantum mechanics presents is that the subatomic universe is wave-like as well as particle-like. Reality consists of probability rather than certainty. In that sense, there is nothing in the subatomic world that is "material" as we usually define the word. There is nothing to grasp and measure the way we grasp and measure things in our everyday, macro-world reality. Also, interacting with an electron or any other subatomic particle means interacting with something whose history isn't known. The particle may have even survived from the initial Big Bang. Remember, there is a finite probability that it could have come from anywhere.

Particle physics isn't a pure study of interaction with BTBB reality, since in particle physics both time and matter are involved. Yet in the transition immediately after the Big Bang, matter had just been freed, and that matter would have had the properties that quantum mechanics describes. We don't know exactly how matter developed immediately at the Bang, so we can't say how interaction with subatomic particles relates to interaction with BTBB reality.

However, quantum mechanics offers strange predictions that hint of timelessness, nonmateriality, and universality. Thus, it offers a glimpse of interaction with BTBB reality.

Quantum mechanics demonstrates that the universe is far more homogeneous and accessible than we may have thought (universality), and it even offers its own form of timelessness. An electron can be in our hand one moment and in another galaxy the next. There's timelessness in that. In a way we're all part of the rest of the universe and the rest of the universe is part of us. This unification may be the strongest evidence we have that there is a relationship between ourselves and BTBB reality. If that interpretation is valid, then there is a bond of all reality both before and after the Big Bang.

We've brought up a lot of physics in this chapter, much of it difficult to accept. It's a big plateful, but there is no requirement for the physics to be understood to follow the arguments of this book. All that needs to be noted is that there is physical, measurable evidence in our everyday reality that hints at the timelessness, nonmateriality, and universality of interaction with BTBB reality. This conclusion becomes even more dramatic if we move on from subatomic particles to a more fundamental aspect of nature— light.

Chapter 9
A Comparison to Interacting with Light

Humans—and other species that have sight—interact with light in a seamless fashion. Although we can play tricks on ourselves with light, we have no difficulty using light itself. We know how to work with light, and, using lenses and mirrors, we know how to make light work for us. On the other hand, knowing that light's characteristics make time relative and include a factor that relates energy and matter presents a challenge for any analysis. Moreover, light's properties contain all the mystery that quantum mechanics suggests. This means that the hints of timelessness, nonmateriality, and universality that were noted in the last chapter apply to light as well. In addition, light offers another intriguing opportunity. Since it travels as fast as it does, our relationship to it mimics, in a sense, our relationship with the Big Bang. Because we're so used to interacting with light, we may be able to use light to anticipate how an interaction with BTBB reality might work. That is, we can consider ourselves with respect to light in the same manner that we consider ourselves with respect to the Big Bang. This is an exciting possibility. It means that by examining our interactions with photons, we'll gain more confidence in searching for interactions with BTBB reality.

At least we're all familiar with light. Even a blind person recognizes that there's a sense available to humanity that interacts with light and presents an image of reality that conveys shape, color, and proximity. Humans exploit this sense to the fullest, using light for measurement, design, and separation. Light defines our physical reality. Anything we come in contact with either absorbs or reflects light. Light can be reflected with mirrors, intensified by a magnifying lens, and even reinforced to act as a laser. It can be split into its components to yield the magnificence of a rainbow.

As far as our bodies are concerned, however, light is inert. Unless it's so bright that it hurts to look at it or focused so that it creates enough heat to burn, we can interact with light without any noticeable effect. Light isn't threatening like some other physical phenomena. When we interact with photons, normally it's a calm, easy interaction. We measure, we admire, and we communicate, using photons effortlessly and freely. We bear witness using light, enjoy spectacular vistas, and, perhaps most important, use light for sharing information.

Light has always been a critical human tool, but modern technology has extended that use even further. We now use light in everything from movie projectors to optical fiber communication. We stimulate photons to study their response, and we collide photons to examine the consequences. We use photons not only to define our reality but to analyze it.

Light can only travel at one velocity, 186,282 miles per second—thus, the relativity of time. If it's coming from a source going half that velocity toward a target, 93,141 miles per second, it still goes to the target at 186,282 mps. If the source is going away from the target at 93,141 mps, light still has the velocity of 186,282 mps toward the target. This odd characteristic applies to all electromagnetic energy. Light is merely that form of electromagnetic energy that humans can see.

Let's broaden our viewpoint and think of the total role that light plays. We now know that light means more to our reality than merely providing sensual information for our eyesight. Physical circumstance affects light. For instance, a large mass bends light, which in turn distorts time, the mass playing a role similar to velocity. Once light is affected in this manner, all reality is affected. And, significantly, an atomic clock will physically measure the time change. Time **really does** slow down. Light isn't merely a tool with a defined speed. Such a definition implies that if we found another tool, that is, another way to communicate at an infinite speed, time wouldn't be relative. Light is fundamental to our reality, and, therefore, it defines our reality.

The effect of physical circumstance on light is important, since it might imply restrictions on our capability to experience BTBB interaction. We will expand on the effect physical circumstance has on light later in this chapter, but, for now, we'll concentrate on the implications of the similarity between light and BTBB reality.

Light is basic to measuring time; that's a proven fact. We interact with light constantly, yet the difference in timeframes between ourselves and photons is similar to the difference in timeframes between ATB and BTBB realities. Our interaction with light defines the pace of our clocks and also defines our interaction with the universe as we know it. Light's fundamentality, therefore, proves to be critical. If our relationship to light is similar to our relationship to BTBB reality, then any interaction with BTBB reality may be interacting with a fundamental aspect of reality as well.

If we are actually interacting with a fundamental aspect of reality, think of the power any interaction with BTBB reality would have. The problem is confirming the hypothesis. Unfortunately, we can only make the comparison and cannot conclude for a fact that we actually are having such an interaction. Nevertheless, we know that we have the capability to freely interact with light, and that certainty leads to the following implication. That is, if we can freely interact with a form of pure energy whose particles— photons—travel at the speed of light, we should be able to interact with another reality, the BTBB reality, that has the same timeframe difference. Of course, BTBB reality too would have to be embedded in our own reality. But if it is, and if we interact with it, we would have that similar fundamental interaction.

Remember, both light and BTBB reality are essentially energy. Also, we easily interact with light that's embedded in ATB reality.

Since ATB reality came from BTBB reality, BTBB reality could be embedded in ATB reality as well. Thus, the complexity of interacting with ATB reality and BTBB reality in that sense is similar.

Now let's get back to physical circumstance. Note the fleeting aspect of our interactions with light. That is, any interaction with photons is very temporary. For instance, when we use a tape measure to measure a room, the photons come in contact with our eyes at the moment we make the measurement. Theoretically, before they do so, photons and humans judge each other as having a much slower time. At the point of seeing, however, when the photon and human realities intersect with one another, our times agree. But the moment is just that, a moment.

Mass is still another consideration, one that's already been mentioned. With mass there's the similarity again as in our relationship with light and with BTBB reality. We have mass, but the photon doesn't. And we have mass, but BTBB reality doesn't. Although mass may be a significant factor, it doesn't prevent interaction with light, even though the interaction is as fleeting as it is. So too, mass shouldn't prevent interaction with BTBB reality. However, there's a significant difference between interacting with light and interacting with BTBB reality when mass is taken into account. Light is part of ATB reality and ATB reality contains matter. The significance of matter is important, since all matter, the matter of the universe ATB, bends light. That is, matter distorts ATB reality. It distorts our interaction with light. As noted above, mass bends light.

Of course, the bending of light occurs on many levels. Mankind has known about this phenomenon—albeit in a different sense—long before Albert Einstein, long before the discipline of physics. For instance, ancient spear hunters had to take the fact that water bends light into consideration or they wouldn't get any fish. Later mankind learned that a curved lens redirects light, refocusing it to allow eyeglasses to function or to concentrate energy that can ignite paper on a sunny day. A mirror can make a person look thin or fat, short or tall. Twentieth-century fun houses used to charge for the service.

The effect of mass that Einstein postulated, however, is far more fundamental. He found that mass changes the shape of space and deflects photons from the path they would have otherwise taken. So, just as mass distorts space and changes the rate of time, so too may it distort BTBB reality. Trying to understand the distortion would be mind-boggling, but distortion must be taken into account when comparing BTBB interaction to human interaction with light.

Light is complicated and challenging, and there's no way that it can be fully detailed here. All of its complex characteristics, coupled with the fact that light and time are intertwined, indicate how difficult it is to compare interaction with light to interaction with BTBB reality. Yet interacting with light is one of the most natural aspects of life, and this interaction proves that we can interact with a reality that is pure energy and travels at the speed of light, the same attributes that interaction with BTBB reality has. A simple extension of that logic predicts that interaction with BTBB reality is feasible.

Human interaction with light enhances the possibility of BTBB interaction. Our experience with light offers not only the capability to show that humans have the capacity to easily and naturally interact with BTBB reality, but at the same time it shows how complex and intricate such an interaction would be. Defining the parameters of the interaction is challenging and imprecise, based on light's almost incomprehensible characteristics. Yet, as already noted, we use light effortlessly. It's natural for human beings to measure and evaluate reality using light. If we can interact with light in the fashion we do, we should be able to interact with a reality that has those same features. That in itself should be encouraging enough for us to pursue the unlikely possibility that we do interact with all of reality, including both the reality ATB and the reality BTBB.

A skeptic may argue that simply interacting with light is so natural, so fundamental, that it contributes nothing to the possibility of interaction with BTBB reality. If timelessness is so critical, they might ask, why isn't there timelessness with light? The answer that timelessness is in the instantaneity of the experience of seeing might satisfy the query, but that answer might raise a more fundamental question. Do photons have a different timeframe than we do? Photons travel at the speed of light, but is their timeframe actually different than a human witness's timeframe?

So a basic question is: when we observe photons, when we have that interaction, was their timeframe different than ours before the actual interaction? That is, have humans been able to make measurements proving that fast-moving particles have a slower time? Do we actually observe this time difference in our reality, or is it merely inferred by the Special Theory of Relativity and unproven?

Chapter 10
Time's Variability Influencing Everyday Reality

The question is: does a time discrepancy actually exist within our everyday, ATB reality for particles going at a high speed? True, we've already discussed the relativity of time at length, but just how universally does relativity apply? Remember, our universe consists of the reality that was born from the Big Bang. Its timeframe—the pace of its time—in this reality should be constant. Time on earth is the same as time in another galaxy. When scientists determine the age of a rock found on the moon, they're using earth's clock as a reference. The age of a rock on the moon should be the same as the age of a similar rock on earth. So would time for fast moving particles like photons within our reality be relative? If it isn't, then all of the conjecture in the previous chapter is futile, since there wouldn't be any time discrepancy between humans and the photons they observe.

Of course this can't be true. Such a difference would violate Einstein's Special Theory of Relativity, and scientists have used this very issue to verify that Einstein was correct. They've measured particles that have different timeframes. The time of moving particles is as relative to ours as the time of that fast rocket ship heading for another galaxy. Einstein's theory has been confirmed over and over.

An example of such a confirmation involves the particle known as the muon. Muons are subatomic particles created when atomic rays from space collide with the earth's upper atmosphere. The muon has a half-life, the average time for half of its population to disappear, of 2.2×10^{-6} seconds. That is, half of any group of muons under consideration will disappear in that time. They just don't last very long. And since their half-life is so short, the newly created muons can't travel much further than 2100 feet. Obviously none of these can reach the earth's surface.

Yet they do!

The reason they can reach earth's surface is that they travel close to the speed of light. So their time must be different than earth's time; their clock must be slower than earth's clock. Thus their half-life as they speed to the earth's surface appears longer than measurements made on the earth would indicate. When we observe a muon that has arrived at the earth's surface from a collision with our atmosphere, therefore, we're interacting with a particle whose clock has run differently from our own. Also, the time difference approaches the one which exists between our ATB reality and the BTBB reality. With a muon, not only do we have proof that fast-moving particles have a slower time, but we have another example to use when comparing the relationship between ourselves and BTBB reality.

Remember, a muon created by collision with our atmosphere can't slow down and still arrive on the earth's surface. If it did slow down, its time would speed up, and its half-life would dictate disintegration before it got here. If it sped up, it could get here, but then it would be traveling faster than light, which contradicts any known possibility. Thus, the only fair conclusion is that its time slows down in order for it to arrive. If we were riding on a muon, we'd be like the astronauts returning from a voyage—we'd age less than anyone on earth. So too the muon ages less, which in the muon's case allows it to survive. Note that the interaction (our ATB measurement of the existence of the muon) can only occur **after** the muon has reached the earth's surface. If we were on the muon, we couldn't communicate with the earth in a reasonable sense. We can't interact with anything going that fast during its journey and experience the same reality as our own. Once we join the muon and take an impossible ride, we're in another reality with a different time. Thus, when we measure the muon's half-life of 2.2×10^{-6} seconds, we do so by monitoring it from creation to extinction in our timeframe. Then again, if we were on the muon as it sped along in its slower timeframe, its half-life would still be 2.2×10^{-6} seconds. We'd have the same experience in either case. The muon's half-life doesn't change, but its timeframe does.

This important point must be kept in mind when we consider differences in the speed of time and is worth repeating. The difference in the speed of time only surfaces when the two timeframes interact, and one of the timeframes is used for the measurement. The muon's half-life is constant. It's the muon's time that's measurably different, and that measurement only shows up when observing the muon after it has sped along on a high-speed trip. Any conjecture concerning time relativity, therefore, must be prefaced with a clear definition of how the interaction is occurring for the measurement.

Note that we're the ones who have exploded away from the reality Before The Big Bang, so we're the ones moving at a high speed. Our "half-life" is, say, seventy years. (For a strict definition of "half-life," 70 years isn't applicable, but let's assume that it is for this argument.) Our clock is slower than a BTBB clock, so our 70-year half-life would appear to be longer when observed by a BTBB observer, whose clock runs much faster.

If we slowed down to meet a BTBB observer, our clock would speed up, and those 70 years would dwindle to nothing in our former ATB clock's time. Similarly, an agreement that our "half-life" is seventy-years applies only if the measurement occurs in either the BTBB timeframe or the ATB timeframe, not from a viewpoint within the BTBB timeframe looking onto the ATB timeframe or vice versa. As it is, the 70 years of our time passes by much more slowly when viewed in the BTBB timeframe. However, if our reality "meets" an embedded BTBB reality, we do the equivalent of slowing down to meet with a BTBB observer, and our clock would speed up dramatically. We'd be like the muon slowing down to meet the earth's timeframe. If it did so, it's time would speed up. The muon's clock would go so much faster, in fact, that it wouldn't arrive.

The understanding of exactly how time changes is critical when identifying interaction with BTBB reality. Timelessness, nonmateriality, and universality can only be used as criteria for BTBB interaction when the interaction involves our ATB reality meeting the BTBB reality. It is similar to the interaction that takes place when the muon meets us at the earth's surface. The muon's half-life is so short that it could never have made the journey from its collision to the earth's surface unless it had a different timeframe. When we measure its half-life as 2.2×10^{-6} seconds, we must conclude that the muon's clock was slower as it sped toward the earth. Similarly, when we see a drastic time differential in our human experience, it is a clue that that experience is actually interaction with another timeframe. If we experience a seemingly impossible amount of change in a very small segment of time, the interaction must be with a faster reality such as the BTBB timeframe.

The muon gives us a good example of how two different timeframes create two different realities. In one timeframe the particle exists, and in another it can't possibly exist. The difference in the situation is the particle's velocity and, therefore, its timeframe. The comparison of its time with our reality demonstrates how our interaction with BTBB reality could also be considered. One timeframe would allow one set of circumstances while another would allow some other set of circumstances.

Normally we only consider interaction with the ATB reality timeframe, so as far as we're concerned that is the only timeframe that matters. But other timeframes **may** matter. True, just as muons wouldn't affect us if they didn't reach the earth's surface, so too BTBB reality won't affect us unless we have the interaction that allows our two timeframes to intersect. But if we have that interaction, BTBB reality affects us, just as the reality of the muon does.

Particle physics proves to be a valuable tool when considering interaction with BTBB reality. The physics is so astounding it suggests that we should question our usual interpretation of reality. There appears to be more there, and that "more" could include the elements that suggest interaction with BTBB reality.

Another aspect of particle physics is also intriguing. Scientists have discovered particles that are "entangled." That is, particles exist such that, if one is affected, another one will be affected as well. What's astounding is that the pairs don't have to be adjacent to be entangled. One particle can be on the earth and the other on the moon, or even in another galaxy, and the coupling is still there. The coupling means that the particles are somehow "adjacent" even though they aren't physically close. This finding further strengthens the argument for the unity of the universe. Though some galaxies are millions of light-years away, somehow they're intertwined with us. There are those who may judge this observation as wishful thinking, but the facts don't lie. Just as the gravity of the sun affects earth's orbit, so too may other bodies affect our reality in ways we don't understand. (I'm not advocating astrology here, but I am noting the implication that these properties suggest about the universality of our reality.)

Particle physics is a difficult subject to discuss, and it is certainly difficult to understand, even when the mathematics is included in the explanation. The complexity can make the non-intuitive conclusions almost impossible to accept.

Quantum mechanics uses probability to explain the behavior of particles. Einstein had great difficulty with that idea. The macro-universe doesn't operate on probability, so why should the micro-universe? Perhaps what appears to be chance is actually defined by laws of physics that we don't understand. Whether the behavior is due to chance or not doesn't matter for the purposes of this discussion, however. What counts is that we can demonstrably interact with subatomic particles, and, when their timeframe is different from our own, we can measure that difference and explain the phenomenon accordingly.

Earlier the promise was that this would be a book about people, not about physics. However, to explain total human reality—total human experience—there needs to be an understanding of all human interaction. To accomplish this, we're looking for evidence of interaction with a reality that existed Before The Big Bang. We promised we wouldn't be satisfied to have discovered that interaction unless we could measure it, and it had the criteria of timelessness, nonmateriality, and universality that we defined. A study of particle physics provides a good beginning. It shows that our reality, as manifested in particle physics, contains interactions that hint at meeting the criteria for interaction with BTBB reality. Though not strictly matching the criteria that the interaction be nonmaterial, subatomic particles—material—behave in an undefined manner that is ambiguously material enough to confirm that possible interaction. The word "particle" itself may be misleading, since often particles have wavelike properties that resemble the pure energy that would meet the criteria.

To establish a meaningful application to everyday life, however, we must bring BTBB interaction to a human level. The odd characteristics of particle physics aren't obvious until studied in a laboratory, and such a study isn't common to human experience. The key to pursuing BTBB interaction on a personal level, however, is to put behind all the prejudices and convictions that earth's clock and matter have pounded into us. A study of subatomic physics not only helps us do that, but it also demonstrates that mankind has the capability and the capacity to recognize interaction with timeframes other than our own. So with this rather superficial background in particle physics, let's begin to apply what we've learned to our everyday lives.

Chapter 11
Looking Internally for Interaction

Humans interact with photons, electrons, and other particles all the time. To understand and to apply this interaction to daily routines, mankind has developed tools that track, respond, and provide us with a visual image. Whether with an electric motor, a television set, a magnetic resonance imager, or an electron microscope, we use particle physics to our advantage whenever possible.

We're made up of particles. Our minds use chemical and electrical dynamics to process thought. If particle physics hints at interaction with BTBB reality, and we use particle physics as a tool, and we ourselves are made up of these same particles, then it should be reasonable to suppose that some form of interaction with BTBB reality could come from within ourselves. In that case we would no longer have the hindrance of another interface before the interaction could occur. Of course, the risk of finding interaction with BTBB reality within ourselves is that personal opinion might creep into any discovery, so sufficient safeguards must be built in to avoid that risk. Our discovery must be measurable and tangible. It can't be mystical or theoretical.

We're physical. Our bodies are material, so, by definition, any human interaction can't be nonmaterial. Our very makeup disallows timelessness, nonmateriality, and, since we're all unique, universality. So within ourselves there can't be pure interaction with BTBB reality. Even thought, which has elements of timelessness and nonmateriality, is physical. It requires brainpower. Consciousness isn't possible if the brain isn't functioning. The brain is material, and it uses time as a reference when considering interaction with any reality. The only possibility for progress in our quest is if there is more to thought than brainpower. That is, if we find a nonmaterial element to thought, we might find the interaction we're seeking.

To find that element let's consider another aspect of our existence—walking. To walk we require far more than leg power. We must have balance, a sense of direction, and a purpose, even if the purpose is as trivial as exercise, enjoyment, or escaping from an irritating situation. Of course, humans aren't the only creatures that walk. Most earthly creatures have some mode of self-transportation, so in that sense they too have purpose and direction. For many animals the purpose is survival, although that's often coupled with some curiosity. Humans may take self-transportation to a higher level, but essentially it's similar for all creatures. The point, however, is that there is more to human movement than merely leg power, just as there's more to fish movement than wiggling a rear fin. So, is there more to thought than physically thinking?

There very well could be. Note that although thoughts require some physical life of the brain, they can extend beyond human experience. Brand new ideas keep coming. Thoughts, even under the most dire of circumstances, continue to surface. Our imagination can run wild, taking our thought process in directions totally different from ones that current physical conditions suggest. Thus, although the brain is physical, its function can transcend immediate physical stimuli.

The human brain's capabilities exceed even that. Humans can concentrate on the theoretical and develop ideas far beyond anything that mankind has ever physically witnessed. Albert Einstein's ideas were eventually confirmed experimentally, but, in some cases, not until many years later. If the brain is purely physical, how could it conjure up a purely nonphysical concept?

How could Einstein discover that time is relative if none of his experience indicated that it was? In fact, his personal, physical experience must have insisted that time couldn't be relative, that it must be absolute. Einstein had to violate his own experience to come up with a revolutionary idea, and he did it solely by thinking. His thought process went beyond the physical brainpower that responds to physical stimuli. Thus, Einstein's thoughts took on a life of their own. True, he used physical reality and mathematics as a basis for these revolutionary ideas, but the point here is that he took that physical reality to another level, a level that contradicted his own physical experience. He reviewed the same evidence that his predecessors had and thought of a new solution to a problem.

Einstein used his physical brain to make his judgment, and that brain consisted of the same particles that we earlier concluded showed hints of interaction with BTBB reality. In this case the interaction was within himself. It wasn't external, so his brush with BTBB reality was far more meaningful than merely observing particle physics behavior.

This phenomenon, creating an original, abstract idea, occurs frequently. All revolutions begin with an idea, whether the revolution is political, social, religious, or scientific. The idea excites and motivates people. The idea convinces followers who build on it. A strong movement may eventually emerge, and huge changes occur, all due to a thought that may have been quietly posted by a timid, though original, thinker.

The fact that humans can consider the abstract and work with it distinguishes us from the rest of earth's creatures. We can work with the nonphysical and make deductions from mere contemplation. Thoughts themselves can extend beyond the physical.

But how can a concept that has no more reality than a well thought-out opinion be as real as an athlete's diving catch of a football? Especially if that football is about to hit the ground in the opponent's end zone when the player's team is five points behind and only one second remains on the clock. Are abstract thoughts just as real? If so, is it because they meet the criteria of human logic, or must they be confirmed through experimentation to be considered real? Was Einstein right when he said that mass bends light, or was he right only when scientists confirmed by observation that it did so? The issue, therefore, is: if thoughts aren't necessarily connected to the normal physical experience, how do we know when an idea is valid?

Of course, abstract ideas go beyond the scientific. In science an idea is proven valid if it's verified physically. However, there are other venues where abstract ideas take hold and are defended to the death, even though these ideas cannot be proven physically. Proponents of an idea, which may range from the mystical to the political, may convince a whole society that it is valid. That society may even wage war to spread the concept and define anyone who chooses to ignore or oppose the idea as an enemy who must be destroyed. Advocates insist that the taking of lives isn't murder if it's done in defense of their view. It's justice. But is the idea valid? That question remains open.

All this quandary originates in thought. So, although thought comes from the brain, which is physical, thought has a real, nonphysical aspect as well. Perhaps exploring the dichotomy—the tangible and the abstract—within the thought process will lead to finding the interaction we're looking for. In any case, it's a good starting point for discovering what drives mankind and defines our existence.

Of course, many will argue that there is nothing new in questioning whether an abstract idea is valid. That quandary has been an issue for as long as there's been recorded history. This argument is correct. This quest isn't new, because we're not looking for anything new, not really. We're only trying to explain what we already experience. The point here is that the human mind pursues the abstract, and the abstract can extend beyond the physical. Therefore, the thought process extends beyond the physical. But when it does so, validity becomes an issue.

This book is faced with the task of verifying an abstract idea as well. When we find what we're looking for, however, it will be what we have defined, not some opinionated, mystical answer based on theory that is founded on more opinion and more mysticism. We will demonstrate the validity of our claim. Of course, there are bound to be advocates and opponents of the result, but hopefully we will avoid the polarization that some concepts foster. We're searching for interaction with a reality that's been in existence longer than our universe. Looking for such interaction within ourselves is not only exciting, but it may explain many of the so-called mysteries that mankind has identified in its quest to understand life and the universe.

Moving the search from the external world of particle physics to the internal world of human thought is a big jump, the only connection being that we're made up of the same material that makes up the rest of the universe. But that connection is significant. What we know to be a fact for particle physics may also play a significant role in our thought process.

BTBB reality preceded us, and in that way we're embedded in it. How, we don't know. Our time is consecutive and our space is dimensional. Spacetime means slices of reality in the ATB universe. We consistently deal with those slices and react within those slices. However, although spacetime is a real concept, it doesn't apply to everyday reality in the routine sense that we define that reality. We can speak in spacetime terms, but in actuality we function in a sequential timeframe that is based on the calendar and the clock.

Yet there may be a domain where all spacetime comes together. Just as in our everyday reality, where electrons can be next to us one moment and anywhere else in the universe the next, and just as entangled particles affect each other even if separated by a solar system, so too does that very same unification apply when we interact with BTBB reality.

Looking within ourselves for this interaction leads to exciting possibilities. Let me repeat that this book will not lead to any outlandish or sensational revelations. If we find interaction with BTBB reality within ourselves, it will be in particular instances and it won't be pure. Nevertheless, interaction with the two realities, BBTB and ATB, no matter what we find, will explain our total reality so much better than trying to explain it only with ATB evidence.

I suppose that many readers, perhaps most, think that if there was a Big Bang, it happened so long ago that it doesn't matter, much less apply to us now. These skeptical individuals may concede that we're the result of some explosion, but they would maintain that we've moved on. Any reality that existed before the explosion is gone, kaput. Live for the future. Forget all this nostalgic nonsense and move on.

On the other hand, mystics may see opportunity here, since the implications are mind-boggling and sensational and not easily explained. The mystics may require a certain mantra and/or some incense before they agree on a role for BTBB reality, but they could work with it. I mean no disrespect. Mystics can be sincere, hard-working, selfless people, who honestly believe in a higher order. However, that higher order is usually in enough of a fog to require "faith" to fully comprehend it. Mystics might embrace the idea of interaction with BTBB reality, but more than likely they wouldn't accept any explanation that left out the mysterious.

In our task we must keep the viewpoints of both the skeptics and the mystics in mind. Healthy skepticism assures that we don't jump to unexplained conclusions. Yet a successful pursuit of our endeavor requires an almost mystic faith that we'll find this interaction with BTBB reality. All the while, however, we will insist that the result be observable and measurable. So we continue our quest. Will we find the interaction we've seen on an imperfect level within particle physics in a more perfect level within ourselves?

I repeat: particle physics has shown that within our ATB reality there are instances that have the characteristics that meet the criteria for interaction with BTBB reality. Our minds operate using those same particles. The human thought process, therefore, may be the most fruitful place to find interaction with BTBB reality within ourselves.

Chapter 12
In Thought

Before we explore human thought, let's revisit the caveats of our search for BTBB reality first defined in Chapter 4. They are: (1) we won't experience an actual physical interaction with BTBB reality as we do with ATB reality; (2) any interaction will be imperfect and have a physical component to it; (3) there isn't a "place" where BTBB reality resides; and (4) confirmation that the interaction is actual will be that BTBB interaction is the only satisfactory and complete explanation for the experience.

One can easily see how two of these caveats apply when considering human thought. Information that feeds our mind and nurtures thought comes from our senses, and our senses are tuned to interact with everyday, ATB reality. Therefore, if there is any interaction with BTBB reality within ourselves, it would have an ATB component. Thus, our physical being plays a critical role in the interaction which precludes that interaction from being pure. Further, to experience BTBB reality, it would have to be within us already. Any external reality would be ATB reality and not BTBB reality. Thus, the BTBB interaction would be with a built-in, pervasive reality that is embedded in us, not in some other "place."

This observation isn't as outrageous as it may first appear. We are as much a part of the universe as any other component. We're made up of the same matter, and that matter has a history that extends all the way back to the Big Bang. Since our very being is the resultant of the process that encompasses our reality, there could very well be something inside us that originated prior to that momentous event.

Our beings are complicated and the complexity adds yet another dimension to our quest, so we must proceed cautiously. Nevertheless, human thought, despite being fed by the senses, can operate on an abstract level. It can have an amazing array of experiences. With that in mind, we'll investigate human thought.

To begin with, let's make an important distinction. Humans appear to have a unique trait within their thought. This trait goes beyond their ability to reason, to use tools, or to modify the reality in front of them. Besides those abilities, humans can imagine, predict, and occasionally sense with undefined "feelings" a reality that goes beyond what their senses are currently grasping. Whether it's déjà vu, premonition, or "gut feeling," individuals often appear to have an intuition about their circumstances. This capability doesn't include mental telepathy or the host of other scams that con artists have invented to dupe the dupable, but it does include the ability to make a leap in judgment that doesn't necessarily follow from known facts.

If intuition is the aspect of human thought that's interaction with BTBB reality, however, we have a huge challenge. We can't measure intuition. And even if we could quantify its existence, we wouldn't be positive that some subconscious experience wasn't influencing our judgment. Therefore, we will not include intuitive thought in our pursuit.

That said, our challenge is still formidable. The process of human thought is arguably the most complicated aspect of human existence. There are both electrical and chemical elements in the human brain, and the brain's methods of processing information have yet to be fully understood. A healthy brain is active, vibrant, and creative. Déjà vu, intuition, and the rest not withstanding, a functioning mind is physically healthy, and it uses its capability in exciting and surprising ways.

But how? What is its physical function? Inside the brain there are electrons, the same electrons that have a finite probability to be anywhere else in the universe later on. (Of course, when we consider brain function, we're considering billions and billions of electrons, so the probabilities are so ridiculously tiny that using quantum mechanics as the reason for losing one's mind isn't reasonable, despite any desperate claim of a struggling physics student.) Since electrons in the human mind give it the potential of universality with the rest of the universe, that universality presents the possibility that we may find the interaction we're looking for in thought.

Timelessness also surfaces when considering thought. Just how long does a thought take? Ever try to measure it? "I've got to think about it for awhile," a spouse might say. No doubt the response is often a delaying tactic to avoid a confrontation, but it may be a sincere desire to mull things over, to weigh the consequences, and consider the pros and cons. In that case, the time is measurable, possibly even lengthy. However, thinking about a problem too long can be unproductive. We can become mired in thoughts that continually repeat themselves, preventing new ideas from entering into the calculation. A lengthy thought process is not necessarily a productive one.

Conversely, there are occasions when thoughts **are** timeless. They seemingly arrive out of the blue, almost instantaneously. A new insight to an old problem may appear that is so original that its author is astounded, delighted, and even in awe as to where the idea came from. The question becomes whether the insight was the result of days or even years of pondering, or does this fresh view of reality arrive like a lightening bolt? Are concepts born subconsciously over time, and is the timelessness of thought an illusion? Or is the experience real timelessness?

With proper equipment scientists have made attempts to measure brain activity, but to couple that measurement with the quality of the activity is difficult, if not impossible. There are too many variables in the thought process, too many interruptions, too many detours following the wrong train of thought to make an accurate measurement of how long the thought took to germinate and then blossom. Novelists can write books in weeks or years depending on how quickly the thoughts come. The actual speed of creative thought is immeasurable. While brain activity itself may be measurable, defining the quality and actual usability of its output complicates any measurement. Nevertheless, there are cases when a new idea is groundbreaking and surprising. At the time the idea surfaces, it's timeless and, as such, offers hope for identifying interaction with BTBB.

We've found universality and timelessness in thought. What about nonmateriality? The brain is physical. Electrons move about in our brain. Remember, however, these are the same electrons that are both wave-like and particle-like. This dual characteristic has all the complicated and micro-material aspects that quantum mechanics implies. Also, thought can be concerned with the abstract and have minimal material content, adding another nonmaterial element to the process.

If I could perch on a fast-moving electron in my brain, my time would slow down, and if the electron was going fast enough, I would—in my new time—have years to watch my brain perform a simple task. From my body's viewpoint, however, the speeding electron's clock would be the slower one. Remember, from a relativistic point of view, when the electron looks at the body, the body is moving. Thus the body is the one with all the time on its hands. But when the body looks at the electron, the electron is the one that's moving and having the slower time. The only instance when the clocks would agree would be when the electron's reality intersects with the rest of the brain's reality. Is that when thought is born?

Of course it takes many electrons to create a thought. All the billions of them in a human brain constantly interact with the brain's host, so trying to analyze the brain by considering the comparative timeframes of any one electron is run-amuck oversimplification. However, we've already noted the ambiguity of particle physics when it comes to pinning down any particular electron. Electrons may be material, but they aren't the material of the macro-universe. For this discussion, the relevant point is that there are occasions when the brain functions in a seemingly timeless manner, and its components have both universal and nonmaterial elements as identified in quantum mechanics. It is on the occasions where the thought process appears timeless that we may find interaction with BTBB reality.

But when does such an occasion occur? Is it only when we have a particularly original insight? Thoughts range from the practical—"I need a coat," or "Where's this crazy book going?" — to the impractical—"I could have made that catch!" Both practical and impractical thoughts can occur within all categories of thought. Let's look at just one category—imagination—examine it and see if there's evidence of interaction with BTBB reality. We'll delve more deeply into imagination later on. However, perhaps just a cursory view can help us find occasions where imaginative thought meets our criteria for interaction with BTBB reality.

Although imagination can be very impractical at times, when it's controlled it can create marvelous works of music, visual art, stage spectaculars and lyrical poetry. Let's consider one product of the imagination—the novel.

A novelist retreats into a world of make-believe that takes over the author's life while writing and, subsequently, takes over a reader's life while reading. In this dreamy world there may be sexual fantasy, heart throbbing violence, slapstick humor, and warm, human relationships, all thrown together at the author's whim. If the novelist's art is practiced well, the experience becomes so "real" that readers have the same emotions they would have had if the experience had been real. This pretend world gives the reader entertainment, the author a livelihood, and satisfies the author's craving for creativity. All of this is done with the symbolism of the written word. The only physical interaction the reader has is eyeing the words and turning the page.

During the reading of a well-written novel, time appears to be suspended. If the author has chosen to set the story in the sixteenth century, the reader readily accepts this new timeframe with only a few passages describing the lifestyle of that period. If the author decides to create a scene with a sixteenth century child fighting with a parent over TV time, the reader will accept even that, provided, of course, that the reader now believes the story to be science fiction.

Ever since Homer—and probably long before—mankind has relished a good story. No matter how outrageous or impossible a story is, if it's presented well, it will find an audience. Technology has helped with the process. From the invention of the printing press onwards, stories have become easier and easier to disseminate. Further, each generation has made its contribution to story telling. Society sees value in the free thought and randomness of an entertaining story. Good stories play in any culture. They're universal.

Timeless, universal, and with minimum physical input, stories conceived by the imagination closely meet the criteria we've established for finding interaction with BTBB reality. Moreover, we can choose how often we have the experience, simply by choosing how often we absorb something from the imagination of another person. Therefore, these opportunities for interaction with BTBB reality aren't necessarily rare.

The issue becomes quality. The greater the degree of timelessness and universality and the less the amount of material input, the higher the quality of the experience. While reading a good story may stimulate our imagination and help us understand our humanity, any interaction with BTBB reality is at a very low level when the three criteria are considered. And even though the criteria are met to some degree, we still must address the question of whether this activity is **actual** BTBB interaction. After all, reading for entertainment may be ATB interaction, having the same value as participating in an athletic event.

Don't be disappointed, though. With just a cursory look at imagination we found possibilities of the interaction we're looking for. We use our imagination in many different ways, and other uses may give us a higher quality interaction and a better understanding of BTBB reality. Moreover, the imagination is but one aspect. There are other forms of thought that offer potential for meeting our criteria as well. Thus, we will continue our pursuit for interaction with BTBB reality by probing the thought process more deeply.

Chapter 13
In the Abstract Conclusion

Human beings are a complete unit. Although we tend to divide ourselves up into tendencies, attributes, and components, in the end we're one being that can't be separated into parts. The thought process is no different. It involves our complete self, not just the mind. Not only do our senses feed the thought process, but our whole history does as well. When we think, our complete being provides input. If we sometimes interact with BTBB reality in our thought process, it means that our complete being interacts. Whether in the imagination or in some other aspect of consciousness, the interaction still will be **within** our complete self. So we'll look at various aspects of thought, including imagination, dreaming, humor, and storytelling, keeping in mind that wherever we find the interaction it means human interaction, not just one facet of humanity. Finally, remember that we've already decided that because we're physical and live in our everyday timeframe, the interaction won't ever be perfect, so we must look for the closest match that we can find. With that, let's consider the thought process as a whole.

Since thought requires time, the temptation may be to assume it could never meet the criteria for BTBB interaction. For example, depending upon the complexity of a problem, a mathematician could work on a solution for years. Also, the thought process relies on the brain, on its ability to process information and eventually resolve what needs resolution. Both time and brain matter count in thought. When analyzing a problem, time may drag while we go over the premises again and again, looking for a solution. We can become so stymied that the solution never comes.

The thought process as a whole may not be any better example of interaction with BTBB reality than reading a great book, since all the physical responses and emotions that go into experiencing a story can also go into thinking through a problem to its solution. However, before we get too pessimistic, let's analyze the problem-solving process. This process can be subdivided into three components: analyzing the premise, weighing options, and drawing a conclusion. We'll look into each of these components.

Immediately we notice that the first two aspects, analyzing the premise and weighing the options, take time. When we analyze a premise, we consider the experiences we've already had with reality. If we've learned that pretty women are sweet, we'll act on that bit of information until it's proved incorrect. After meeting the first nasty, conniving female with a beautiful smile and an attractive figure, we'll decide that the premise is wrong, and from then on we'll take that experience into account. In fact, we may overcompensate and penalize an innocent creature just because of one unpleasant experience. Unfortunately, judgments can be biased. Only more experience will provide the wisdom that tells us that looks have little to do with character. A gorgeous creature sitting across a conference table might then be evaluated on her views and talent rather than her physical appearance.

In analyzing a premise, we use preconceptions that may or may not be valid. To form an opinion or to make a decision, we use the experiences we've had with everyday reality. If we didn't, we'd be refusing to learn from past mistakes. To operate most effectively we must keep an open mind, yet temper that open mind with an accurate interpretation of prior experience. We also take into account the experiences of others, learning to weigh that information against the reputation of those who've related it. New experiences will modify our opinion, and, as long as we keep ourselves open to new information, we can fight any prejudice that might creep in. The important point is that, as we make the analysis, time passes, even though we may not notice it. All of our history plays a role, and it takes time to review that history, even if the review is at the subconscious level.

Weighing options is no different. To begin, we consider each option in the light of previous knowledge that we acquired either directly through personal experience or indirectly through schooling, advice, or even hearsay. Should we vacation at the beach or in the mountains? Should we live near our work for a shorter commute or further away for a more defined escape? Should we cook out on the grill, microwave a gourmet prepared meal, make something from scratch, or eat out? Should we marry someone of the same faith, or is the person more important than his or her religion? We face all these options, and we consider them one after another as we live our lives.

Weighing options can be instantaneous or the process can take months if the design of a building is at stake, and there has to be a strength analysis made for each shape considered. Good thinking weighs all the options available in a thorough and objective manner. Legal argument, where ideally the final decision is based solely on law, is a good example. In law, if the judgment is arrived at properly, it is made in a precise, defined manner with only the laws, their past interpretation, and the constitution used as guidelines.

Whether the issue is common or extraordinary, whether weighing options is methodical and time consuming or instantaneous, a judgment is based on experiences of everyday reality. Since both the analysis of the premise and weighing options can be time consuming and are embedded in everyday reality, both offer little possibility insofar as interaction with BTBB reality is concerned.

However, reaching a conclusion is different.

An argument can be made that reaching a conclusion is the lengthiest part of the process, but actually the whole time is taken up in analyzing premises and weighing the options. Timing the actual reaching of the conclusion is much trickier. After months or even years of work, a conclusion may appear so suddenly and even spontaneously that we wonder why it took so long to arrive. There's an insight that abruptly gives us a light-bulb moment which brings together many thoughts and refocuses them into an original way of considering an issue. We suddenly reach an astounding new conclusion.

A good example is Einstein's discovery of the Special Theory of Relativity. After a lengthy period of work, Einstein thought of looking at the problem in a new way, and that insight revolutionized physics. Simply by realizing that light travels at 186,000 miles per second and never varies, Einstein concluded that time is a variable in our reality.

What was Einstein's state of mind when he reached that conclusion? Was he thinking deliberately? Although a young man, he'd been working on the problem for some time. That in itself would appear to be enough to eliminate the timelessness criteria. Yet a new insight suddenly appeared in a timeless fashion. Where did it come from? It certainly didn't come from everyday experience, so it couldn't have resulted from merely weighing accepted options. Einstein had come up with a new option, a revolutionary thought that disproved part of what scientists had understood to be fact up until then. There was a new interpretation of reality involved, a different interpretation that scientists had not even considered.

Maybe that was it! Maybe at that point Einstein actually interacted with BTBB reality.

Think of that moment and compare it to the criteria of interaction with BTBB reality that we've defined. The breakthrough was tremendous, yet it came in a flash. That is, all of Einstein's thoughts came together and produced an astounding new idea in that one moment of insight. That's our definition of timelessness. Further, it was pure concept and thus nonmaterial. The conclusion was revolutionary, but it wasn't super-human. Any educated person can review the conclusion and agree with it without possessing a unique capability. Even though Einstein's talent and persistence gave him a special insight, his uniqueness was the degree to which he used his deductive reasoning powers. Significantly, the ability to deduce applies to all normal people. It's universal.

Now for the tough part. Although this example meets our criteria, is it actually BTBB interaction? Could it be something else? Could it simply be ATB interaction, the brain's left side interacting with the right side in an original and exciting manner? Did Einstein's brain possess a unique talent that was able to function in a special way?

Trying to explain this phenomenon using the criteria of ATB reality, however, raises many questions. Why was the jump so sudden? Why weren't there timely steps in the process to make the transition smooth? Einstein had spent years contemplating the problem, yet there was a sudden insight. It came after a more or less casual conversation he'd had with an acquaintance. Einstein went home and pondered the conversation, when, almost instantaneously, the thought came.

Why didn't Einstein require more physical evidence to make his logic-jump? How was he able to use a universal capability in such an unusual way? Why didn't he limit himself to the physics of the issue at hand, instead of expanding his horizons into the intriguing question of whether time is relative or absolute? What in ATB reality presents the BTBB criteria that Einstein used for his insight? That is, where is timelessness, nonmateriality, and universality evident in ATB reality? Which is more reasonable to conclude: that these properties are part of ATB reality, but somehow hidden so they're not self-evident in the physical world, **or** that these properties are part of another reality, the BTBB reality? These questions must be kept in mind when making any judgment. Considering what BTBB reality is and how our interacting with it would take place suggests that using it to explain an experience such as Einstein's is far more reasonable than fabricating some unknown aberration of ATB reality for an explanation.

As already noted, Einstein's experience may have been intense and revolutionary, but it wasn't unique. Its distinction was in the **profoundness** of the conclusion's meaning. We all experience conclusions that are both significant and surprising. While the degree of significance may vary, the conclusions can affect our whole life, and they may seem to come from nowhere, a complete surprise that is life-changing. Who hasn't had some type of experience where a sudden significant thought has made an impact? These moments may not be as revolutionary as Einstein's, of course, but they do exist.

Again, there are arguments against Einstein's conclusion being an interaction with another reality, much less BTBB reality. Einstein used his mind to develop his thoughts, so they weren't fully nonmaterial. Also, abstract ideas like the one Einstein had are not always universally applicable. For example, when general relativity's ideas are extended to the sub-atomic world, there is a breakdown in its explanations. Note, however, that while the experience of having such an idea must be universal to meet the criteria for BTBB interaction, the idea itself doesn't necessarily have to be universally applicable. Finally, the time for the conclusion to be reached may have been infinitesimal compared to the time Einstein spent understanding the premise and weighing the options, but it was a finite time.

Therefore, a fair judgment would be that although Einstein's experience can't be properly explained as ATB interaction, it wasn't pure interaction with BTBB reality either. On the other hand, we've emphasized that pure interaction isn't possible. What we're looking for are the times when mankind comes closest. Einstein's discovery could be one such example, despite its imperfections.

Finally, humans respond to an important conclusion in different ways, and their responses might cloud any interaction that's taken place with BTBB reality. It's critical to note that any human emotion, such as elation or celebration, cannot be included in the interaction we're investigating. The response occurs afterwards, when the thinker is back to interacting with everyday reality, not at the point of making the conclusion. We'll take a closer look at the actual point of making the conclusion in the next chapter. For now, the important thing to note is that a conclusion like the one Einstein made essentially meets our criteria of timelessness, nonmateriality, and universality and cannot be reasonably explained using ATB reality.

Chapter 14
Examples by Reaching a Conclusion

Any BTBB reality that exists must be pervasive. It can't be a separate entity, since all of our ATB reality sprang from it. Similarly, for us to have the capability to interact with BTBB reality, the BTBB reality must be within ourselves. Furthermore, any such interaction must be unlike interaction with the physical, ATB world. Drawing a conclusion, particularly on an abstract subject with broad implications, meets that condition, since finding an ATB interaction with the same characteristics isn't possible. True, ATB interactions, such as winning a world championship, can bring equivalent celebration, but the equivalence ends there. And, as we've already noted, human emotion following the realization of a profound insight isn't part of the BTBB experience. So let's look more closely at drawing a conclusion to pinpoint how the interaction takes place.

Human behavior doesn't fundamentally change over time. Modern science may be able to explain human behavior more accurately than it's been explained in the past, but the behavior itself transcends generations. No matter what we determine to be interaction with BTBB reality, it isn't going to be new to mankind.

It shouldn't be surprising then that when finding an aspect of the human condition that essentially meets the criteria, the actual discovery isn't earth shattering. In fact, now that we've made the discovery, we may feel rather blasé. There's nothing sensational about someone drawing a conclusion. Even when reaching a revolutionary conclusion, we celebrate the conclusion itself rather than the actual experience of reaching it. We may admire the author, but we don't see the accomplishment as interaction with another reality. A great mental discovery like Einstein's signifies a superior intellect with amazing insight but not super-human capability. We all reach important, abstract conclusions—perhaps not as significant as Einstein's—and some of us may do so frequently. They're eureka moments, but they're not the abnormal or strange moments that we may have imagined interaction with BTBB reality would be. The important point is that BTBB interaction explains the why and how of that moment better than other ATB options might. Now let's delve into how we arrive at that eureka moment.

First, the actual point of reaching a conclusion is unique. It's the instant that all of our previous thoughts on a topic come together. Often at the precise moment that we reach the conclusion, we don't even realize we've done so. We conclude and then we act, and frequently the two occur so closely that they blur together. The actual moment of decision may go unnoticed. Whether we're cognizant of the experience or not, however, we make a judgment.

What is the judgment based on? Does it follow the evidence exactly, or does some new element enter into the equation? We may naively label it as intuition or sixth sense, but the judgment may use information that, up to that point, hadn't been included in the weighing of the facts. This is especially true if the subject is theoretical rather than concrete. The more abstract and profound the subject is, the more we're removed from our everyday experiences. Our thought process may soar, and somehow we're transported into a reality where time, matter, and other physical stimuli don't count. As we arrive at a critical conclusion, we do so without emotion. Tears, laughter, or anger may come later, as a reaction to a startling discovery, but emotions don't play a role in the actual concluding process. We're in a state of suspension, grasping all the information we have, looking for possibilities and then, perhaps abruptly, arriving at a conclusion.

Another aspect of the conclusion process surfaces when considering someone else's discovery. That too may lead to interaction with BTBB reality if the conclusion is abstract and revolutionary enough. For example, think about Einstein's groundbreaking conclusion from the point of view of someone who's just heard of it. If that person is skeptical and non-scientific, pondering the relativity of time may be difficult at first. Trying to accept the idea that time is variable and depends upon the relative velocity of two points of view often leaves a novice student in disbelief. If life teaches us anything, it's that time is constant. Besides, even if the theoretical argument is accepted and a mathematical demonstration is understood, the skeptic may decide that since relativity applies only to speeds humans will never reach, there's no practical application to everyday reality. Then some agitator writes a book like this, claiming interaction with BTBB reality, which only confirms the skeptic's judgment that relativity isn't practical.

An open-minded reader might rethink the issue, however. Since the relativity of time is proven, objective thought might allow that there is nothing that forces relativity to be considered only from the point of view of earthbound humans. Why not look at relativity from the point of view of Before The Big Bang? After all, there had to be reality then, and there is no absolute rule that that reality doesn't play a role in human existence now. Thus, this objective person—hero in my view—draws the conclusion that the search for interaction with BTBB reality is worthwhile. Perhaps when the hero forms that conclusion, he or she comes close to interacting with BTBB reality.

There are other instances where we have similar breakthrough moments like Einstein's and, therefore, have a type of BTBB interaction. For example, consider a situation that isn't abstract, like convincing a lover to marry. If there's anything that's immersed in ATB reality, it's marriage, since marriage involves living in the everyday world, with its health and sickness, wealth and poverty, happiness and tragedy. However, obtaining the actual commitment for a lifetime of living together often requires using the abstract, the sublime, the romantic, the fanciful, and even the magical.

Let's consider this example from a feminine viewpoint. The young lady loves the young man with all her heart and is convinced that her feelings are reciprocated. Now she's ready to settle down. Yet the young man only infrequently suggests a lifetime commitment, and, when he does so, doesn't offer a proposal. What's the young woman to do? Not hindered by tradition, the young lady isn't too bashful or too proud to initiate the discussion herself, but is unsure of when it would be best to do so. If the moment is wrong, she decides, the attempt might result in more harm than good. So she procrastinates, suggesting with her eyes that he propose, but the suggestion is never noticed. Suddenly it dawns on her that her very absence might be the stimulus that will shock the gentleman into his senses. She becomes quite busy all of a sudden, too busy for their usual rendezvous. The ploy works. The young man realizes his situation and understands there is only one way to solve the problem. A month later, he produces a ring.

At some point in the thought process from when the girl decided to take things into her own hands to when she told her beau that she couldn't spend Saturdays with him, there could have been interaction with BTBB reality. There was the breakthrough moment of a new idea. The exact point of the interaction might be difficult to spot, but there would have been a moment when neither time nor matter applied. It was at that universal moment, the moment when the originator of the thought understood its advantage, that interaction with BTBB reality would have occurred. Refusing company to someone you love goes against logic. Yet the young lady understood that her heartache would be shared and that the response might resolve her issue.

This concrete example might not be as satisfying as a more abstract one. Certainly there isn't the profound jump in logic and the timelessness that an abstract thought can produce. So for another example, let's go back to the abstract and look at working out a mathematical problem.

In the study of mathematics, word problems often present a difficult challenge. For instance, grade school students may be able to add columns of numbers, multiply and divide, and get the correct answer even though they may have to use fingers to help with the calculation. When it comes to word problems, however, those arithmetic skills are merely background, because a word problem challenges the student to put facts into mathematical terms. For instance, if Michael is carrying a dozen eggs in a basket but drops three on the ground, how many are left in the basket? To an adult, that problem is simple subtraction, but to someone who's just learned that three from twelve is nine, the issue isn't so obvious. As problems get more challenging—a motorboat going 12 MPH across a mile-wide river with a 6 MPH current, how long will it take to cross the river to reach the point directly across from the start?—the solution is less obvious, even for an adult.

When we were students many of us gave up on word problems. If we were fortunate enough to have classmates who had the insight to frame a question into mathematical terms, we could use them to find the answer. Few of us bothered to ask these nerds how they arrived at a solution, but if we had we might have gotten an insight into interaction with BTBB reality. Actually, if we had spent more time on the problem ourselves instead of relying on those bright students, we'd have gotten an even better insight into that interaction, because the degree of difficulty enhances the experience. A bright student may arrive at the solution seemingly out of instinct. The answer simply flows out. Those of us who are less fortunate have to work harder, but that doesn't detract from the legitimacy of the BTBB interaction when it occurs.

If possible, try to remember one of those word problems you worked on. You probably don't remember the details, but you may remember a time when you suddenly understood how to solve a problem. You may have been having difficulty with some other mathematical issue rather than a word problem. Or perhaps it was a nonmathematical theory or concept that you simply couldn't understand. Think about the moment that the light-bulb went off. Recall an instance when you finally understood something that had baffled you. That was the moment of BTBB interaction.

The more intense the thought that goes into reaching a conclusion, the more satisfaction there is. I emphasize, however, that the experience we're considering is not the satisfaction, nor the celebration that goes with it. We may experience a sense of power or some other emotion when we make a particularly exciting conclusion. Even if we stifle the self-congratulatory bragging that good manners have taught us to suppress, we'll still have an urge to share our good fortune. However, we'll share the discovery, not the experience of making that discovery.

What's more to the point is that we'll eventually document the conclusion and the logic building up to it, not the experience we had at the moment of reaching the conclusion itself. Nobel prizes are presented for the conclusion not the process, that is, they're awarded for the results of BTBB interaction not for the moment it took place. The moment of conclusion is difficult to define, so we move on.

We are all grounded in everyday reality, and even the most sophisticated academic who's embedded in the theoretical will use everyday reality as a base. When we think in the abstract, however, we're in a special mode. The actual process of drawing a conclusion while in that mode gives us an experience that is timeless and nonmaterial. It's universal in that it's a common ability that many people exercise. Human emotion, the personal celebration of the discovery, comes later, when the significance of the conclusion is realized.

Since BTBB interaction can occur in varying intensities, our interpretation of the experience varies as well. Perhaps it is just as important to be able to identify the experience as it is to have the experience itself. Once we recognize we're interacting with another reality, we can appreciate the value of the interaction. Celebrating a groundbreaking idea is one thing, but realizing that the idea came from something deep inside total reality is another, even more satisfying discovery.

A truly intense experience, one where a revolutionary conclusion surfaces, may only happen once in a lifetime. Also, such an experience usually happens in the first third of life. We've already noted that many great theoretical breakthroughs happen early, when a young mind has absorbed much of what is known about a subject but still is open enough to be capable of developing a fresh perspective.

Reaching a conclusion can allow us to go beyond the experience of interacting with ATB reality, as Einstein did. However, in every case, everyday reality is the reality that's used as a basis for thought. We're embedded in ATB reality, and there is no escape from that. If we didn't have that restriction, if we could escape from our body and its demands, we might be able to experience interaction with BTBB reality in a more perfect form. While there are those who may claim such capability, those claims aren't verifiable. We won't know and can't know pure interaction with BTBB reality as long as we're stuck in this timeframe. However, that restriction doesn't disqualify the approximations we do experience.

One could argue that for average human beings the experience of a breakthrough-conclusion is infrequent, so infrequent that our day-to-day experiences are essentially only interaction with ATB reality. Thus, although an interaction with BTBB reality may be significant, it plays a minimal role in life because it is so rare. However, in analyzing other aspects of human thought, we will find additional examples of interaction. While they may be of lower quality, they are still important.

Chapter 15
In Imagination

Thought is wide-ranging and varied. Abstract thought requires analyzing concepts that can range from the fanciful to the theoretical to the practical. However, although abstract thought occurs in many different areas, it requires the use of imagination. That is, to think in the abstract, the mind must modify the immediate reality that the senses detect and change it into a theoretical reality that the mind imagines. Perhaps the next place to look for interaction with BTBB reality, therefore, is in imagination. However, please keep in mind that any interaction will not be pure. ATB reality is always mixed in with BTBB reality. When we discuss imagination—or any other experience—we will consider it within total reality which includes **both** BTBB **and** ATB reality. So when we speak of imagination we're speaking about interaction with a total reality having a possible BTBB component.

With that in mind let's look at imagination as a whole. What is its purpose? Why modify a reality that we're embedded in to create another reality that may place more demands on our existence? Why is the demand for this alternative reality so strong at times that some individuals resort to artificial stimulation such as hallucinogenic drugs in a futile attempt to force their minds into some other reality? This chapter will begin our exploration for the answers to those questions.

The imagination compels us to examine the limitations of physical stimuli and insists that we consider other alternatives. Imagination is the tendency, perhaps even the propensity, to interact with another reality that is beyond the one enveloping us physically. The imagination itself may be the strongest evidence we have that there's a natural inclination to interact with BTBB reality.

Searching within imaginative thought for BTBB interaction presents a new challenge, however. Imaginative thought can play tricks that may result in intentional or subconscious misdirection. That is, the reality isn't what its author claims. This possibility means that with imaginative thought the author may or may not have had interaction with BTBB reality. The interaction may seem to meet the criteria, but it does so only because of an author's erroneous claim. Therefore, for the purposes of this discussion, we must define what we call "genuineness." Then we will give examples of the use of the imagination (1) where we know the use **is false**, (2) where it **might be** genuine but there's no certainty that it is, and finally (3) where we **know** it is genuine. Once we have a clear understanding of the three possibilities, we can proceed. Note that since the human mind is capable of imagining almost anything, the question isn't whether the concept itself is being imagined. The question is whether the claims of the imagination are genuine. So what do we mean by asking that question? What is genuineness when we're considering interaction with BTBB reality?

If we're really interacting with BTBB reality, we are actually having that interaction. It's not a fantasy that we misrepresent. We are interacting with a reality that existed before the Big Bang. The interaction meets our criteria of timelessness, nonmateriality, and universality and cannot be fully explained by ATB interaction. While we may not use the terms of this book to relate our experience, the experience is real. It's genuine. We may interpret the experience incorrectly or misstate what we experienced, but we actually had the experience. Our claim is sincere and, to us, as actual as any interaction with ATB reality. Others may disagree with our result or distrust our interpretation, but the interaction occurred. That's what we mean as genuine. In the case of the imagination, we'll restrict our examination to cases where we **know** the experience is genuine and take up cases where we're not sure in part three.

Even with the above definition, however, there may not be a full appreciation of the implications of genuineness. We'll give some examples in this chapter to further clarify the issue. So don't be concerned if the concept is still a bit fuzzy. For now, **the definition of genuineness is that the experience meets our criteria, cannot be adequately explained by ATB interaction, and is actual, not made up or purposely distorted.**

So let's look at some examples. Consider palmistry and mental telepathy. Are these genuine interaction with BTBB reality?

Practitioners who claim they can read the future or read minds also claim they have a special ability that few people possess. They have to, because the rest of us know we can't do either. This claim violates the universality criteria of BTBB interaction, and therefore those uses of the imagination cannot be interaction with BTBB reality. They're probably not sincere either. The only way the experiences could be valid is if the perpetrator proved the capability beyond a doubt using accepted and verifiable criteria. To date no one has done that. So-called seers either make predictions so vague that they can't be confirmed, or they make observations that could apply in any situation, or they downright cheat, drawing on information they pretend they don't have.

Say, however, that a fortuneteller admits it's all a trick. Then does the activity include BTBB interaction? The perpetrator is using a skill to fool people, and that skill uses the imagination. The reality that is presented is a modified reality that, theoretically, we could all experience if we learned the skill. Thus, in that sense, the experience is universal. There may be timelessness and nonmateriality in the skill as well, and the whole event may not be adequately explained by ATB reality. Thus, even palmistry and mental telepathy could include some genuine interaction with BTBB reality, but **only** if the experience is portrayed as a trick.

Next, let's look at the second instance, that is, an example that we think may be genuine, but we don't know for sure. This type of example is more difficult to analyze. The example here will be simplistic, but several important aspects of genuineness will surface, so the simplicity helps. Unfortunately, the example is also quite lengthy. Please be patient and follow along, because if you do, later on you'll see how the arguments apply to more complicated situations.

Okay, let's ponder something basic, like shelter. Humans require shelter for protection against the weather—the cold of winter and the heat of summer. ATB reality insists that we have a minimum amount of shelter, enough to protect ourselves from the elements with sufficient space for freedom of movement to perform our daily tasks. Different people require different amounts of room in their shelters. Those who share common areas with others might only need a small cell for themselves. Monks, for example, need little personal space. On the other hand, leaders require more space, since they must occasionally gather those they're leading for communication.

Although there is some variation in the amount of space required for each individual, ATB reality defines a minimum, below which a person can't function properly. There also is a maximum. For example, a pantry must be relatively close to the kitchen, and the kitchen must be close enough to the eating area so that hot food can be transported without getting cold. Washrooms must be within easy walking range from anywhere within a dwelling. While there is a flexible range for the distance between all of these facilities, there **is** a limit. If space is too tight or too expansive, those living within that space simply can't function.

The question then becomes what is a **reasonable** range? To answer that question one must use the imagination. For example, ATB reality may dictate a maximum distance between critical facilities, but when is a house simply too big? Even though its occupants may be able to function adequately, is a huge mansion reasonable? Some may call it opinion and others may call it taste, but there is a point where a dwelling can be considered to be too expansive and a waste of resources. Perhaps the structure simply doesn't fit in with the neighborhood, or perhaps it takes up too much land, or perhaps it's considered ugly. Whatever the reason, a problem is defined and action is demanded. Of course, people are free to "waste" if they choose, but some may decide that such waste is bad for society and attempt to legislate against it. Activists may create an organization that rails against such "materialistic" excess. The organization may grow into a "Movement" where new rules are defined which represent a modification of the ATB reality at hand. The rules come from the movement's leaders, who must use something to create them. That something is the imagination.

Let's say the "Movement" requires its followers to limit the size of any dwelling they inhabit, and it sets a limit of 500 sq. ft. per person. Anything more is defined as excessive. The point is, that to establish the limit, the leadership had to use imagination; it had to **imagine** a reasonable limit. Similarly, to agree with the ruling, the followers had to imagine that the requirement was reasonable.

Some people (especially members of a "Movement" whose goals are more serious than those in the admittedly trivial example above) might question the use of the word "imagination" in the development of their policies. To label their rules as products of the imagination could be considered insensitive. Members of the "Movement," especially if it is religious, might think such a designation mocks its beliefs and is blasphemous. The beliefs are **real** reality as far as they are concerned. However, no offense is intended.

Remember, by definition, any modification of ATB reality using human thought is imaginative no matter where the thought comes from. The thought may have been divinely inspired, but it's a thought that extends beyond the physical circumstances of ATB reality, and, using our definition, the new reality is imaginative. The product is **not** the same ATB reality that all earthly creatures experience. Thus, whether the movement is religious, such as St. Francis of Assisi's, or nonreligious, like Karl Marx's communism, it focuses on a modification of ATB reality, a new definition of reality that its leaders define using imagination, and which its followers accept through their imagination.

However, what is this "new" reality? Is it simply an ATB modification of ATB reality, or is it something fundamentally different which results from interacting with BTBB reality?

We can answer this question by applying the criteria we've established. Was the use of the imagination timeless? Perhaps. The final number may have been arrived at abruptly based on a leader's sudden insight. Of course, on the other hand, it may not have been. Was it nonmaterial? Certainly space is material, but the idea of limiting space isn't physical. The idea that there should be a limit is nonmaterial. Is it universal? There's nothing special or out-of-the-ordinary in establishing limits. It's common practice and a perfectly normal activity that humans perform. Even other species establish territorial limits. So, yes, it's universal. Can this decision be explained by ATB reality alone? ATB reality doesn't require such a limit. The only maximum limit ATB reality requires is that necessary facilities be close enough to allow their use. So to proclaim some other limit suggests going beyond ATB reality. Therefore, couldn't using imagination to get the idea for setting a limit essentially be interaction with BTBB reality?

Perhaps yes, perhaps, no. The experience of coming up with the idea to set some sort of limit is an internal one, and we cannot read someone's mind. The reason for the uncertainty lies in the claim of timelessness. The leader may claim to have had some type of timeless, a-ha moment of inspiration, but there is no way to prove that claim. We simply don't know enough about the experience. If the author of the rule chooses to enhance our understanding by explaining the experience in detail, we could make a better judgment, but only if we knew that the explanation accurately reflected the experience. To know that actual BTBB interaction was involved, we must know that the interaction was not contrived, that the imagination really did present a timeless moment of insight. The problem is that we don't know if the explanation is sincere.

Let's put the issue into the terms of genuineness. First, we've already conceded that we don't know what type of interaction took place. The rule may have been purely an ATB interaction with no genuine BTBB component. We don't know for certain; we can only make an estimate. And, since we have no certainty, our judgment about the genuineness could be correct, or it could be in error. We can't know.

There is another issue here as well. In addition to determining whether the experience was genuine BTBB interaction, there is also the issue of interpretation. That is, a person may genuinely interact with BTBB reality but misinterpret the interaction. The result could be just as serious as misinterpreting interaction with ATB reality. In the latter case, the result might be a fall resulting in injury or even loss of life. In the square footage example, the rule could be faulty whether or not there had been interaction with BTBB reality. If there had been BTBB interaction and it was interpreted incorrectly, the result would be an incorrect rule. So even genuine BTBB interaction does not guarantee success.

This brings us to the third possibility—a case where we know the use of the imagination is genuine. There are a number of such cases, and we'll be exploring several examples in subsequent chapters. For now though let's look at advertising. Within that discipline there is the potential to use the imagination in such a way that it meets all the criteria we've established for interaction with BTBB reality, and the use cannot always be completely explained using ATB reality.

In advertising there has to be a hook—a clever tagline, image, or trademark—often used as an anchor. There is BTBB interaction in the development of that anchor. The anchor, of course, can range from poor to average to great, but whatever it is, its birth is quick. While there may be a number of hooks evaluated, we're not considering the process of deciding which one will work. Here, we're only concentrating on the birth of the idea. That idea will show up in a timeless manner. For some cases the insight may be so trivial that it's not worth mentioning, but in other cases the insight is so dramatic that its timelessness shouts out. The better the anchor, the greater the insight, and the more intense the timeless moment. In any case, at all levels of advertising, a timeless moment is genuinely there. The anchor itself may have a nonmaterial element—a catchy tune or poetic slogan—or it may be all ATB reality, but in either case, its discovery is cerebral and nonmaterial. Finally, although developing a product anchor may require a certain talent, it is a talent that we all have after a fashion. So, to some degree, we see timelessness, nonmateriality, and universality in the use of the imagination for developing advertising programs.

The product or cause is usually ATB reality, and ATB reality, therefore, is the reason for the advertising program. However, while ATB reality might explain why the program exists, it can't fully explain the elements of the advertising program. The program is often designed to appeal on several different levels, and some of those levels may have nothing to do with the ATB reality of the product or the cause. A clever and effective image may be quite contradictory to normal ATB reality. Thus, advertising programs cannot be totally explained by ATB reality.

In addition to meeting the requirements for BTBB interaction, advertising is always genuine. Advertising methods use imagination to first hook and then inform the public. Advertising may have questionable ethics or a sleaziness about it that makes us uncomfortable, but it is what it claims to be—an enticement or information that the advertiser wishes to convey in order to inform or convince an audience. And to accomplish that, any timeless moment within that process will be genuine. It's in that use of the imagination that there is BTBB interaction.

In a way, advertising is merely a subset of storytelling that we'll explore in Chapter 18. For now, however, just note that in using our imagination to interact with BTBB reality there are times when there is no ambiguity as to genuineness. Advertising is just one of those cases.

Thus, we've given examples of the three degrees of genuineness when using imagination for interacting with BTBB reality. Palmistry is an example where there is no universality and, therefore, no interaction with BTBB reality; social rules are examples where genuineness is ambiguous; and advertising is an example of genuine BTBB interaction. There are other, perhaps less controversial, examples of genuine uses of the imagination which we'll be touching on in subsequent chapters. For now, all that's important is that we understand genuineness and how it applies to the imagination using BTBB reality.

To drive the point about the significance of genuineness home, let's look again at invalid claims and compare them to valid claims of the use of the imagination. Consider the story of a man who has great strength, has the ability to fly, and can see through solid surfaces.

If the story is portrayed as fiction, this new reality will be accepted as such and enjoyed, based on the rules that the author's imagination incorporates into the story. Superman doesn't present a problem to his audience as long as he's portrayed as fiction. However, what if an author proclaimed such a story to be nonfiction? Think of someone claiming to know of a person who actually had Superman's powers. Unless that person could be seen performing these amazing feats, it would be impossible to verify the author's claim.

Yet there are claims of a similar nature made all the time. People claim to hear voices, to be able to communicate with the dead, to have the capability of seeing into the future, or to understand the coordination of human activity with celestial bodies. Some claim they can cure diseases, make the crippled walk, or eliminate pain. Others claim that they have communicated with extraterrestrial creatures. These claims are most likely a scam, but are any of them legitimate? A complication is that some individuals actually become empowered with the new "reality" they believe they've found. The crippled **do** walk. Why? What's actually happened?

The power of mental suggestion is a reality in itself. Scientific experiments testing new medications must be run using a placebo, since the mere act of taking a pill and thinking it may be a cure can create a positive effect. This result demonstrates that imaginative modification of ATB reality can happen and can be genuine, temporarily at least.

Suggestion is a method of selling, but the ability to sell an idea doesn't necessarily mean that the idea is what it is represented to be. It might be actually what's claimed and good, but it could also be false and bad. Or it could be false but good or actual but bad. Authors of false, poor ideas of dubious origin can be great salespersons, but that only makes the judgment of the validity of the claim more difficult. Tricksters can create an intriguing and promising reality that is just far enough removed from ATB reality so as to be difficult to disprove. And, as already noted, sometimes the fakery actually convinces individuals and changes their ATB reality. We don't know if these tricksters are using BTBB reality or not, but in either case we know enough to be wary of them. Any claim of a great insight must be considered carefully.

We're creative, thinking beings. We continually make choices, and in those choices we use our best judgment to distinguish the valid from the invalid. Nevertheless, history teaches us that human beings often make incorrect judgments in that regard, and the results can be catastrophic. Identifying fakery is easy once the falsehood is exposed, but discovering it beforehand is often impossible if the purveyor of the falsehood is talented enough. Con artists have been around since society began, and they'll keep coming as long as mankind's imagination stays active.

These are some of the problems we face when we attempt to find interaction with BTBB reality using the imagination. Although imagination offers much potential, it also presents complications that are difficult to overcome.

But there is an answer. There are aspects of the imagination that are universally accepted as genuine and provide even better examples than advertising. These aspects are as legitimate as a novel and just as easy to define as valid. That is, a novel claims to be fiction, so it can say almost anything. As long as we find it entertaining, we can enjoy it as a legitimate use of the imagination. So too we can use other products of the imagination that make different claims. In our search for using imagination to find interaction with BTBB reality, we will ignore the claims of pundits, unscrupulous politicians, and other purveyors of unverifiable "truths" and concentrate on aspects of imagination that are demonstrably genuine.

Later we'll address ways to confirm whether other types of interaction with BTBB reality are genuine. We can apply those methods to some extent when addressing validity for the imagination as a whole. For now, however, we'll concentrate on imaginative experiences where we can agree that the claims of those having the experience are genuine. For example, there is a form of human imagination that transcends everyday reality but is innocent and very real. This use of human imagination is exactly what the purveyor claims it to be. It has the added value of not containing a history of restrictions and preconditions that humans tend to load onto imaginative experiences. This aspect of imagination is the very special imagination of a child.

Chapter 16
In Childhood Imagination

Remember the carefree days of childhood? If there were worries, they usually came from restrictions imposed by adults, convinced that they knew better. Perhaps they did, but if those restrictions had been removed, who knows what the freedom might have brought? Those days were days of innocence. While we may have played tricks or even connived to gain an advantage, the tricks were fun and entertaining. We enjoyed ourselves and let our imaginations run wild. If left alone, our imaginations could create marvelous worlds of fantasy that included heroes and conquerors overcoming nasty villains and horrific obstacles. Children can slay dragons, fly through the air like Spiderman, fight the bad guys of the old West, survive intergalactic travel, battle sea monsters, and do it all before lunch. None of the constraints of adults matter to a child's imagination.

Youthful imagination is genuine. That is, the claim of the experience of a child's imagination is exactly what it is. It's as genuine as anything adult, even though its output may be of no more use than an adult piece of fiction. But the genuineness of childhood imagination is special because of a child's openness. It meets our criteria for interaction with BTBB reality, as does an adult story, but it does so on a new level, since a child is quick in creating and accepting the alternative reality that an imagination can create. A child doesn't require all the props and physical stimulation that adults often do. Adventures come easily and frequently, and, though a child will accept any story easily, much of the freedom lies in the child's **own** imagination.

There is much to learn from a child's imagination. Although adults frequently dismiss the pretend world of their offspring, that pretend world can be very real. It offers the child a chance to compete on equal terms. It offers creativity, surprise, and humor. Children can put a new twist on an old idea, and that new twist is often so original that adults wonder where it came from. Even something as simple as a child using a word in a new way can offer a surprising insight on how to view reality.

A child's imagination provides its own unique perspective on interaction with BTBB reality. Though children use imagination to play, their play not only suspends ATB reality and replaces it with another, but the replacement brings with it a perception of time far different than an adult's.

For youth time drags. Christmas takes forever to arrive even after the month of December begins. While adults rush about trying to prepare for the holidays, children wait...and wait...and wait. Anything anticipated takes forever. Like birthdays. A birthday could come the week before Thanksgiving, yet a child may not even associate the two. Even though Thanksgiving offers a four-day weekend soon after the birthday, to a child, Thanksgiving is Thanksgiving and a birthday is a birthday. The two are separated, unrelated, and, therefore, not nearly as concurrent as they must be for the parents who are busily preparing for the holiday, thinking of what to get their child for the birthday, all the while knowing that Christmas is just over a month away. For the parents this time is frantic. For the child, time drags. It stalls. The birthday finally comes, then the long Thanksgiving weekend, and then there's another eternity until Christmas vacation begins.

Why does youth, with its vivid imagination, think time is so much slower than adults do?

To answer that question requires an understanding of the changes that occur in the pace of time as the years pass. Time appears to accelerate as we age. It takes around a thousand years to reach the magic age of sixteen when eligibility for a driver's license kicks in. The twenty-first birthday with its legal beers takes another few decades. Then, before an eye blinks, it's a fiftieth birthday; the kids have moved out, and the future is fixed. The aging parent may feel lost, impotent, and isolated. That's when time may begin to drag again and continue to drag until a doctor diagnoses a pain as a life-threatening condition. Then time may speed back up, so that death approaches like a charging beast. Time, at least the perception of it, seems to vary. Sometimes it drags and other times it's fleeting. Busyness or health may be factors, but they're not the only ones. What appears to be a very critical factor is the activity of the imagination.

Although our perception of time varies, its measurement is constant. Seasons come and go every year and with them the annual holidays, anniversaries, and other celebrations. Time in our everyday reality only appears to change. Yet why would time give that impression? Why would a youngster with a vivid imagination think time was dragging, while a middle-aged woman who'd just passed menopause wonder how life could have sped by so quickly? The two viewpoints are contradictory.

A day is twenty four hours and it doesn't vary. If twenty four hours appears longer for youth, it's because those hours contain more. That is, the imagination is working for those hours, and while the imagination is active, time isn't important. After all that imagining, a child thinks that days should have passed, but they haven't. Perhaps only a few minutes passed while the youth imagined flying off on an eagle to another land, gliding through the sky toward a lake, and having a meeting with druids on a mysterious island. Now that the youth is finished with the adventure, there is plenty of time to take out the garbage, which is only a short walk out the door, but a dull walk, an unimaginative walk—a walk that takes forever.

A vivid imagination means a more active mind, which, one would think, would make time pass faster. That is, if a strong imagination is more associated with interacting with BTBB reality than a less active one, then youth's time should go faster, not slower. Actually we've just seen that it does. The imagination allowed the child numerous adventures in a short amount of time. So, though the child thought time dragged, it was only because the child was comparing BTBB time to an ATB clock. That is, a child's timeframe is much slower than an adult's, simply because, in the child's cycle, the imagination soars, and the child concludes that time must be flying by. The child has been interacting with BTBB reality with its much quicker timeframe. Compared to that experience, everyday reality appears impossibly slow.

Contrast this with an adult's view. For adults, life is very structured and full of obligation, thus, not enough time. While both the adult and the child may accept that one hour has passed, the child thinks more time should have elapsed, but an adult thinks it should be less. Thus, there's a significant difference in perception between an adult and a child.

However, when adults have extra time to think and reminisce, their sense of time can be similar to that of a child's. That is, after spending time in contemplation, the adult thinks much more time should have passed than actually has. Of course, an adult's more mature mind may be less agile than a youth's, so the contemplation may not be as intense. Nevertheless, when contemplating, adults experience the same time difference as a child. (Note: There is another instance when time appears to go more slowly, and that's when a person is bored. Boredom, however, is simply irritation with the clock's speed. For example, staring at a second hand makes time much slower than when the clock is ignored. Although boredom may not include literally watching a clock, it occurs when doing nothing of interest, which is virtually the same thing. Boredom is essentially an ATB experience that has nothing to do with BTBB interaction.)

Thus, the significance of childhood imagination for our purposes is its perception of time and its genuineness. The other characteristics of childhood imagination, universality and nonmateriality, offer little new insight. That is, children in all cultures use their imagination to play and pretend. Doing so is as universal as making an abstract conclusion. And the child's imagination has the nonmateriality that all thought has. However, childhood time and genuineness are special. In youthful imagination we find the freshness, the brashness, and the timelessness of a clock running infinitely faster than our own.

In addition, childhood imagination is usually free from the scars that everyday reality leaves. Children have the advantage of having thoughts unburdened by the years of experience that fence in adult thoughts with artificial boundaries. Perhaps that's the reason that youthful minds have the most original ideas. Einstein's Special Theory of Relativity, the theory upon which this discussion is based, came when he was in his twenties—twenty-six actually. His General Theory of Relativity came when he was thirty-seven. It's fascinating to realize that for the rest of his life he pursued a universal theory that he never discovered and hasn't been found to this day. While working on that universal theory, he spent considerable effort battling the advocates of quantum mechanics. He too became inhibited by those damnable fences that come with age.

Although youth has the advantage of freshness uninhibited by experience, that characteristic can also be a disadvantage. Wisdom and experience, the very things that build the fences, keep mankind from stepping into minefields and blowing themselves up. Thus, mankind's imagination is a complicated combination that balances two forces, the uninhibited imagination of a child and the tempered imagination of a mature adult more in tune with the complications of ATB reality. While youthful imagination may be less contrived, it is also less practical. On the other hand, mature imagination (more practical but more contrived as well) is more difficult to confirm as genuine.

Wisdom and experience may set firm boundaries for adults, but that doesn't necessarily mean they form a barrier for interaction with BTBB reality. So what is the barrier? Perhaps neurons simply aren't as active in an older mind, and that's why, as already mentioned, mathematicians and other scientists tend to make their discoveries early in their careers. Neurons may move more freely in a younger mind. Perhaps their paths aren't as set. On the other hand, wisdom allows the mature person to evaluate facts more accurately. A wise adult can deny or confirm interaction with BTBB reality more readily than a youngster considering the possibility. Youth jumps to conclusions too easily and may even decide that such interaction, if it occurs, is the only interaction that is worthwhile. Perhaps one aspect of the generational conflict is due to the collision of these two interactions, one with everyday, ATB reality and one with BTBB reality.

Youth with the imagination that's obsessed with finding a more meaningful life, as well as the elderly freeing themselves from the entanglements of life, often insist that there is more to reality than the everyday drudgery of ATB existence. Finding actual interaction with that "more" is a greater challenge than simply proclaiming that the reality exists, but youthful imagination gives us another glimpse of mankind's attempts at the discovery.

We've already mentioned that youthful imagination meets the criteria we've established. There's nothing material about youthful imagination. It may use our ATB reality as a foundation for its reality, but the new reality it creates can transcend any physical laws that ATB reality presents. Youthful imagination certainly is universal. The shrieks heard on a playground sound the same throughout the world. Some of that noise is due to ATB interaction, of course, since most playground equipment fosters that type of interaction. But some of the noise comes from children pretending, imagining all sorts of things. The laughter confirms it. Childhood imagination is the same the world over. But it's the timelessness, as manifested in young people's experience of time, that adds the most to our understanding of interaction with BTBB reality.

Let's try to find another piece of the puzzle by considering a different aspect of human imagination that's also genuine. It has to be, or it won't work. Let's consider humor.

I can anticipate a reader's reaction. Humor? Now he's trying to find interaction with BTBB reality in humor. This guy's got to be kidding!

Chapter 17
In Humor

Few things are more enjoyable than a good laugh. When we laugh, time and matter disappear. A good joke is universal, transcending cultural and generational gaps. The joke uses a common link within humanity's view of reality and identifies a place where that view simply doesn't fit.

We don't measure the quality of a joke by how long we laugh or by some spatial dimension. The joke is judged by its intensity, and the intensity is measured by the degree of suspension it gives us from the here and now. Studies have shown that a good laugh heals, that it reduces blood pressure and refreshes the spirit. A humorless person is difficult to befriend, but a person with a great sense of humor makes friends easily. Who doesn't enjoy a companion who makes you laugh? Women can fall in love and marry for that very reason. "He made me laugh," they'll say, when pressed for the reason they accepted.

Youngsters enjoy humor as well. They use their imaginations for humor just as readily as they do for fantasy. They laugh easily, twisting old ideas into fresh shapes and giving them new life, often while mouthing song lyrics their parents don't understand. Children love to play pranks on the older generation, just as that generation played pranks on the generation before, all of the gamesmanship using humor and imagination to wring out a tiny advantage and level the playing field.

Humor can also bite and have a purpose. Its use can even lead to significant social change. Political cartoonists have sharpened their wits while using that aspect of humor for ages.

The product of humor in ATB reality may be laughter, but humor offers much more. Think of the last time you heard a good joke. For a fleeting moment there was nothing more interesting or relevant than that great punch line, and, if you were lucky, the punch line was so good, so appropriate, and so outlandish that you laughed and laughed. Meanwhile the rest of reality disappeared into the background. Time didn't matter. Troubles, difficulties, even pains were gone. A good joke can do that. Material needs retreat to the background. True, some humor applies only to those who understand subtleties that apply in a given culture or situation, but once those subtleties are understood, the humor is universal. "Laugh and the world laughs with you," is a saying that follows naturally from the certainty that everyone enjoys a good laugh. Humor transcends ATB reality, so much so, that some say it extends life.

What makes humor so special? Why do only humans appreciate a good joke? Dogs wag their tails and seem to laugh, but their wagging is a demonstration of joy rather than mirth. Humor is special, perhaps, because it highlights the futility of human excess. It mocks pomposity, reducing us all to the same level. It makes the unbearable bearable and can make sense of this often nonsensical world. Perhaps that's why we love it.

On the other hand, how can humor teach us about interaction with BTBB reality? Humor certainly doesn't go well with physics. Physicists are known for their quirky ways, which may be the subject of levity for outsiders, but physicists themselves are too absorbed in their train of thought to catch any jocularity in their appearance or demeanor—which, of course, is what makes their visual impression even more humorous.

Not all physics professors fit the image, however. I attended an all-male school where we didn't even have the opposite sex around to encourage us to make jokes. On the other hand, our best professor, who easily was the most challenging, the most learned, and the most instructive, was female. Dr. Rose A. Carney was as organized as the numerical system and as sharp as a laser beam, even though at the time laser technology was in its infancy. Dr. Carney had worked on the Manhattan project while in graduate school before eventually settling in as a professor at an all-male institution. She had a dry sense of humor, and when she used it, it was effective. We'd smile but never laugh out loud, because we were too much on edge, too worried that she'd call us to the blackboard to work out a problem.

Despite the humorlessness of physics, however, humor still can add to our understanding of interaction with BTBB reality. We've already noted humor's timelessness, nonmateriality, and universality, but humor may have some other quality that will give us a deeper insight into interaction with BTBB reality. Let's take a look.

Humor entertains, but so do dramas, thrillers, and mysteries. Yet humorous books seldom do as well as serious ones. Why would that be? Why would a humorous book sell fewer copies than a book that makes its reader cry?

Perhaps humor needs to be more personal. When it comes from a source that we can hear and see, it may be more effective than words off a page. Perhaps humor is better if shared. That may be why a lousy stand-up comedian working a crowd often gets louder laughs than a great joke that's written down. But why would a joke need to be shared to make it more effective? What does that tell us about interaction with BTBB reality?

The answer lies in the way time is suspended with a good joke. When shared with an audience, a joke actually suspends time more effectively as everyone laughs. This ATB reality reaction is contagious, which encourages a deeper appreciation of the humor, suspending time more intensely as we absorb the comedy. In the written word the joke is private, and, though humor may suspend time, the suspension isn't shared. Therefore, it seems less intense. Perhaps this example gives us a deeper clue about the relationship of the two realities. Emotional reaction is ATB reality, and we're embedded in that reality. We can't have any experience where ATB reality is totally unaffected. But for humor, ATB reality enhances rather than subtracts from the BTBB experience.

Immediacy may also play a role. Humor is centered on the here and now, or what is portrayed as the here and now. Experiencing the here and now privately isn't as powerful as when sharing it. When a group has the same experience, the experience reflects off its members, and the reflection builds immediacy.

The U. S. Presidency is a good example of the importance of immediacy in humor. The president is frequently used as a topic for humor, and this is due to humor's most obvious aspect, that which makes it what it is, the exaggeration or misplacement of a slice of reality which magnifies an absurdity. Although few of us can get through a day without doing at least one thing that doesn't fit the circumstance, when the president does something inappropriate, the effect is magnified simply because of the grandeur of the office. A satirist uses those missteps to get a laugh. The audience understands that the account is an exaggeration, but sees the speck of truth in the thought, sees the contradiction and the inappropriateness. So the listeners laugh. Again, whoever occupies the office is the subject of the humor. Note that the humor follows the office, not its occupant. If the satire was focused on a former president doing the very same thing, the results would be less effective. Immediacy intensifies the humor.

This isn't to say that the written word can't be humorous or that humorous books can't be successful. It's just that the immediacy and sociality of the spoken word strengthens the humor. Thus, by considering humor to be an example of interaction with BTBB reality, we've learned that such interaction is strengthened with immediacy. Also, the more concentrated the experience is, the more it is enhanced, and having others share the experience concentrates it. Therefore, BTBB interaction is more effective when it's immediate and shared.

We defined timelessness as the equivalent of a significant jump in circumstance happening in an incredibly short time, a time so short that it's immeasurable. Concentrated, immediate humor meets that definition. The result of humor, however, doesn't. Laughter is ATB reality feedback for a human emotion, as tears are for crying, a loud voice is for anger, and trembling is for fear. Emotions involve biology, and biology is physical. We react to a joke as we react to pain or to an aroma or to an interesting taste. Our reaction involves our senses, and it takes actual time for our senses to see or hear the humor. During the experience, the time it takes to absorb and react is ATB interaction, as is the duration of the laugh. The only BTBB experience here is the moment time appears suspended.

To illustrate this point, consider someone asking us to force a laugh and keep laughing for one minute. We'd find such an exercise tedious and exhausting. Yet when the laughter is in response to a good joke, the same effort becomes pleasurable. If we could grab a moment during the laughter and hold on, thereby freezing that particular point of spacetime, we would, because we'd be enjoying ourselves so thoroughly, and that enjoyment would remain in the present instead of in a memory. Eventually the effect of the humor passes, however, and we return to everyday reality with all the frustrations and disappointments that clutter our lives.

Neither the length of time that we laugh nor the time it takes to tell a joke applies to BTBB interaction. The only thing that's meaningful is the humor itself, and that's what's timeless. We might cherish the memory and even try to share it with a friend, but the timelessness would be gone, and our memories can only approximate, not duplicate, that timelessness.

What was the timelessness like? How can we analyze it? How was it different from normal reality other than our being oblivious to the passage of time?

It isn't that we're unaware of our surroundings during a humorous experience. If a gunshot or explosion interrupted the humor, we'd react immediately, snapping out of any interaction with another reality and into the immediate reality of the explosion, protecting our loved ones and taking cover the best we could. Humor may suspend everyday reality momentarily, but even in that suspension we can react. However, more trivial matters, such as a mosquito buzzing or the sun burning the skin may go unnoticed. One could argue that our ATB reality is suspended only to the degree that the joke allows. A better joke removes us further from ATB reality and requires everyday reality to give us a bigger jolt to regain our attention.

Other ATB reality is subordinated even in our reaction to humor. During laughter we aren't concerned with anything but the humor. What we see isn't important unless the joke is visual. What we hear matters only if it adds to the humor. Any taste or aroma is ignored unless it's so powerful that it brings us back into everyday reality. During laughter, we ignore the sense of touch. Even arthritis sufferers may experience pain relief when the laugh is most intense. Our hands may clap in appreciation, but we don't actually experience them touching each other. All our senses are suspended as we enjoy the bliss of a reality where logic is replaced with a twist of thought that is so unlikely, so outrageous, or so clever that we marvel and laugh, oblivious to the rest of reality.

Though our senses are in suspension, our minds are active. The mind plays an important role in absorbing the information and understanding the irony. Although our senses may absorb the circumstance that leads to humor, they withdraw at the precise moment that the humor—and the new reality—kicks in. It's at this timeless moment that BTBB reality surfaces. We're transported to a new state. Immediately afterwards we will return to ATB reality and react, laughing, laughing, and laughing some more.

Humor meets our criteria, albeit with the exception that when we encounter humor we must laugh; the laughter is involuntary. If we tried to stifle laughter, it would be, if not impossible, certainly uncomfortable. Pure interaction with BTBB reality should be nonphysical. True, our senses are suspended from the physical to the extent that hearing, sight, touch, smell, and taste are subordinated, but it's an ATB subordination, replaced with the ATB reaction—laughter.

This is the same reason that sexual climax can't be considered interaction with BTBB reality. There is nothing more physical than a sexual climax, even though at that moment we might feel transformed. All of our senses play a critical role in the chemistry of the sex act. Try interacting with a partner while avoiding the five senses and see how long that partner sticks around. True, romance is required to get in the mood, and romantic foreplay may hint of BTBB interaction, since it uses imaginative methods that can extend beyond ATB reality. However, the actual sexual climax is purely physical, purely ATB interaction.

To a lesser extent the same issue applies to humor, since laughing itself is physical. That fact doesn't mean that humor can't be considered as an example of interaction with BTBB reality, however, because the BTBB aspect occurs prior to the laughter. Also, as we've noted before, any interaction on the human level won't be perfect, since mankind is embedded in time and matter. Humor plays an important role in our lives, and it must be genuine or it won't be funny. When we laugh we're reacting to a unique interaction with BTBB reality, and that interaction gives us feedback of timelessness, nonmateriality, and universality.

So in humor we find another instance where human experience at least partially meets our criteria. Humor gives us the added insight that social participation adds to the intensity of BTBB interaction, and that immediacy increases its effectiveness. Humor also is helpful for understanding timelessness and the role of human, ATB interaction resulting from that timelessness. In these ways humor adds to our understanding of BTBB interaction. And we're not finished yet. Let's look at some other aspects of the imagination that are just as genuine as humor.

Chapter 18
In Methods of Storytelling

We've looked at two aspects of human imagination that are genuine and partially meet our criteria for interaction with BTBB reality—childhood imagination and humor. Although the experiences weren't as intense as reaching a groundbreaking, abstract conclusion, they gave us new insight into BTBB interaction. Childhood imagination enhances our understanding of the role of time in our lives, and humor shows that using an ATB reaction, laughter, to share a BTBB interaction intensifies the experience.

In Chapter 12 we considered how reading a novel meets our criteria. We saw that since a novel exists mainly for entertainment purposes, the BTBB experience is only one of escape. A play or movie offers a similar, low-quality experience, but there are notable differences, depending on the method of storytelling. So in this chapter we will compare the experiences of watching a movie, attending a play, and reading a novel. In doing so we will acquire new insights into BTBB interaction that are based on the comparison.

Before we begin the comparison let's do a quick overview. The result of all storytelling is mainly ATB interaction. That is, stories use ATB reality, or a recognizable modification of it, to create a situation within ATB reality and work with it. Here we're comparing the BTBB component of the interactions. However, within the comparison we'll be noting the ATB differences that are used in the presentation, and we'll consider the input's effect on the BTBB experience. Remember, the BTBB experience is of low quality, so the intensity of the influence of ATB reality shouldn't be surprising. In our search for BTBB interaction we will identify ATB influences and attempt to subordinate them, so we can concentrate on the aspect of storytelling that increases the probability of BTBB interaction.

Both live theater and movies offer escape and insight into human existence, as does a novel, but the distinction here is that each of these three types of expression requires its own level of active participation. We'll start our comparison by distinguishing the experience of live theater from that of a movie.

No matter how hard it may try, no matter the number of speakers or the size of the screen, a movie simply doesn't provide the depth, intensity, and rhythm of a live production. Dancing is demonstrable proof. One would think that film, with its close ups and varied viewpoints, would make a dance far more intimate and meaningful than witnessing a live performance of the same dance from an upper balcony. But it doesn't. Experiencing live dance is special somehow. It expresses a mood that can't be captured on film. It's like trying to portray the full grandeur of the Grand Canyon in a photograph. The image may be there, but the magnificence isn't. Two dimensional film, even in I-Max, can't capture the mood of a live production, in spite of how spectacular the cinematic gadgetry may be.

The difference between live theater and film goes beyond mere perception and into the basic structure of the experience. To keep the audience's attention, a live performance must use its stage efficiently. Dialogue must be crisp, distinct, and hard-hitting. If there's physical action, it can only play a supporting role. A fight scene is only successful if it's set up properly through dialogue. Musicals have the advantage of offering both music and dance to enhance the story, but the experience only works if the music and dancing inspire and lift the audience into a new reality. In every case, the intensity of the new reality is what's important. If the intensity isn't there, the production will fall flat. Perhaps that's why many musicals—and operas, for that matter—borrow their scripts from classic stories. The music enhances what is already known to be a great tale, substantially increasing the likelihood that the production will be a hit.

An excellent play grabs its audience with intense human conflict and clever mental surprises. Visual gimmicks, such as eye-catching disasters, automobile crashes, or huge explosions, may provide thrills and be exceedingly entertaining, but they don't portray a complete story that has the potential for interaction with BTBB reality. The experience that visual gimmickry does present is essentially an ATB experience.

Live theater must rely on the spoken word and human reactions to those words. Give that restriction to a Hollywood screenwriter and watch the frustration. *Star Wars* wouldn't work. Most movies wouldn't work. Movies need action. Clever dialogue helps, but in itself, it won't find a wide audience. The *Titanic* must sink. Indiana Jones and James Bond must have their death-defying physical adventures. The audience didn't pay good money just to hear people talk. Thus, like theme park attractions, most movies offer an experience primarily with ATB reality, and those movies must be culled out for this discussion.

The comments above are not meant as an attack on the film industry. Movies are entertaining and enjoyable. We're just stating the fact that movies exist to grab the largest audience they can, and that usually requires a physical spectacle. A movie, for the most part, doesn't use audience imagination to the extent that a play or novel does. The exceptions are often box office failures, and Hollywood producers rightly ask what good a great movie is if only a few people see it. Therefore, movies tend to rely on the physical, the spectacular, and the adventurous. That is, they rely on ATB reality, not only to present the story, but for the story itself. There may be some higher level thought buried in these movies, and, if one searches enough, it will surface. But, for most casual viewing, action and adventure dominate.

On the other hand, there are thoughtful movies with abstract themes that probe the human psyche, such as: *The Graduate, My Dinner with Andre*, and films taken from plays such as *Amadeus* and *Who's Afraid of Virginia Woolf?* Finding interaction with BTBB reality is far more likely in movies like these.

Live theater uses dialogue as the primary means for conveying its reality. Since it must rely on ideas rather than action, it must present those ideas in an imaginative and interesting way. That requirement presents a powerful challenge to the playwright and performers, but the dialogue worked so well for Lee Remick in *Wait Until Dark* that sophisticated Broadway audiences screamed out loud at one point. True, a physical threat caused the scream, but that threat was set up with dialogue and timing that made the threat believable as well as surprising. The important distinction between a movie and a play is that in a play the imagination of the audience must be used to create the intensity. A movie, on the other hand, can use spectacular physical images of ATB reality, thus reducing the need for imaginative audience participation to create the same level of escape.

A good play takes its ideas and thrusts them front-and-center, using ideas as the means of escape. If done cleverly enough, our imaginations soar during the production. When the imagination soars—note **not** necessarily when emotion heightens—BTBB interaction is more likely, since using imagination allows us to meet our criteria more than other human responses.

Repeating: the main difference between a movie and live theater is that live theater requires a more active participation from the members of the audience, and that participation involves the use of imagination. Coupled with live theater presenting a more vivid ATB reality than film can, as exemplified in dancing, this characteristic presents a distinction that we can use in our quest for learning more about BTBB interaction.

(Interestingly, a movie's ATB reality is often more spectacular than a play's, even though it isn't as vivid. A movie can portray the scope of a hurricane disaster, whereas live theater cannot. However, although that image is more spectacular, it still is not as vivid as it would be when witnessing an actual storm, as would occur in live theater if it were feasible.)

Watching live theater can involve a timeless experience. This experience is universal as well. Virtually every culture has some form of "stage" production, whether it be ritual or storytelling. And although all stories use everyday reality for props and verbal interchange for discourse, a nonmaterial element involving an abstract idea gives intensity to the experience. Even a love story falls into that category, as it concentrates on the nonmaterial reasons why the pair is attracted to each other. The attraction has to be more than physical, since, although physical beauty is the easiest to portray, it's often the quickest to grow stale, even in a short play. There must be additional issues to hold an audience's attention. Granted, live theater doesn't meet our criteria as cleanly as a novel, where the only physical element is turning the pages and reading the words. Nevertheless, it offers a genuine alternative reality, even though that reality is either fictional or simulated non-fiction.

The way a play's alternative reality is different from a movie's is what offers an opportunity for additional insight into BTBB interaction. A movie with an abstract theme can yield the same timelessness, nonmateriality, and universality that a live production can, but since its ATB reality is less vivid than an actual presence, the experience may not be as fulfilling.

Using an analogy with pure ATB interaction, think of attending an athletic contest. A television production of a football game can present a clearer view of the action than watching it live, yet fans will spend a small fortune to witness a critical game in person. Why is that? It's more than a status symbol, since fans who couldn't care less about social status will still pay the exorbitant amount.

Seeing something live allows a person to witness the total event. And, as a witness, one can choose to concentrate on what one wishes. An eyewitness fan is not a captive of the camera angle. True, cameras can present different angles of the same play, but multiplying the number of angles will not necessarily cover the interest of a particular viewer. Often, the fan simply wants to experience the play and the atmosphere surrounding it as a "whole." Or perhaps the fan's interest lies in watching the reaction of the players who aren't in the game. Cameras seldom are interested in sideline action unless there's a fight. The "whole" is much more than the partial reality that a camera can provide, and the same issue applies when comparing plays and movies.

Thus, when considering BTBB interaction, a play allows a more intense view of the stimulus that sends the imagination into a BTBB interaction. True, the stimulus is ATB reality, but that's true for any BTBB interaction. Even drawing an abstract conclusion is based to some extent on ATB interaction. Repeating for the umpteenth time: no BTBB interaction is perfect. All such interaction requires ATB input of one form or another. The point is that the input in live theater is more effective than in a movie where the audience is a slave to the camera angle, and the reality is portrayed less vividly than in a live production. Yet even though the stimuli are more vivid for a play, a play also requires more input from its audience in order to grasp the story. Movies can provide tiny details and show events separated by space and time that live theater cannot. But while the details may be fewer for a play, the stimulus is stronger and the audience participation is greater.

Now let's compare a play to another form of storytelling—the novel. The written word requires the reader's mind to be much more active than it is when watching a play. In a novel, the reader's imagination must provide the images the author is attempting to portray. No matter how well a book is written, the final picture is that of the reader's, not the author's. The act of reading requires interpretation, concentration, and the necessity of filling in gaps with the reader's own experience. For example, if an American read an African novel, written by a Zulu whose life experience didn't go beyond the local tribe, the American would have a difficult time understanding it. African and tribal references would be meaningless, unless they were explained in terms the American could follow. References to common human traits such as greed, envy, lust, and sloth would come through, but the American would surely miss the cultural implications and subtle tones of the plot. Unless an author presents a story in a manner that references situations that the reader knows, the imaginative capability isn't there, and the story won't work.

An audience watching a Zulu play would have the same difficulty, of course, but there would be many more clues during the performance to tell what was going on, such as facial expression and tone of voice, props and scenery. While the audience would still not have the depth of understanding of a tribal member, it would have help. The people would actually be witnessing the event rather than reading a description of it. In summary, all the play's sensual stimulation of sight and sound presents a far more complete physical reality than the written word, thus limiting the role of the imagination. True, plays don't relate the thoughts of a character as a novel might, but, although thoughts help, they don't restrict the imagination as clearly as the physical environment of a play.

As in reading a novel, there is some subjectivity in witnessing a play, since life's experiences are always measured against the play's story. However, subjectivity has much less of a role for a play than for a novel. The audience watches the play as a group and, for a particular night's performance, members of the audience essentially agree on what they've seen. Some playwrights may purposely present ambiguity to create controversy, if nothing else, but beyond that the audience will understand and accept the reality that the playwright and the director intended.

Movies are even more specific. The camera focuses on important details in close-up shots, so the viewpoint is even more defined than in a play. Movies force an audience to accept a very detailed, elucidated reality, even if it be fantastic or frivolous.

So how do the differences between these types of expression help us understand BTBB interaction? First, keep in mind that they all offer a low-quality interaction. They may free their audience from ATB reality, but that freedom is demonstrably artificial and shallow. Within this low-quality experience, however, there is a distinction. Movies that have an abstract theme may offer interaction with BTBB reality, but they restrict their audiences far more than a play or a novel does. The restriction is the least for a novel, where a reader is free to draw a personal picture based on previous experience. The important point is that although all three present a low-quality interaction with BTBB reality, the one with the most freedom also requires the most effort. Since the novel offers the imagination the most opportunity to express itself, it is also the choice that offers the highest quality of BTBB interaction.

From this discussion we've seen that BTBB interaction requires a personal contribution. When you think about it, that applies to humor as well. The more that's put into it, the funnier the humor will be. The best humor requires the background to understand it. If I made a joke about the relativity of time, it wouldn't be effective unless the listener was familiar with Einstein's theory. A comparison of these three modes of expression demonstrates that the more the effort, the greater the impact of the interaction. So even with a low-quality experience, such as story-telling, we can learn something about interaction with BTBB reality.

Life is a myriad of experiences and most of them are essentially pure interaction with ATB reality. However, we keep identifying other experiences that offer something more and which lead to a more complete understanding of human life. These experiences all meet, in one way or another, the criteria for interaction with BTBB reality. Argue, if you must, that partially meeting the criteria only twists the logic and forces the reader to agree to the premise. But if you accept that argument, there must be some other explanation for the existence of these experiences. I submit that for these latest examples, the reason goes beyond the mere human desire for entertainment. Storytelling's very popularity demonstrates the need to interact with another reality—the BTBB reality..

So the imagination continues to tease, and we're not finished with it yet. It is complicated, varied, and layered. Buried within it may be other, more significant, approximations to interacting with BTBB reality. James Thurber captured imagination at work in a play called *The Secret Life of Walter Mitty*. The play entertains audiences by allowing them to witness Mitty's imagination hard at work. Mitty uses his imagination to escape from everyday, mundane reality. The audience is drawn into Mitty's secret life, reminding them of their own daydreaming. We all dream of bigger things, even though the dreams may be impractical, improbable, or even impossible. Perhaps Thurber portrayed a good example of interacting with BTBB reality when he wrote about Walter Mitty. What about the daydream?

Chapter 19
In Daydreams

Dreams offer a fascinating insight into thought, but only daydreaming qualifies for interaction with BTBB reality. Nighttime dreaming is another issue. Reality gets jumbled in our sleep, events become exaggerated and ridiculous, and we live out disasters and victories that could never occur during our waking hours. We're not in full control when we dream in our sleep. That lack of control allows our brain goes off on its own, and the thoughts we experience seem to be gibberish. Nighttime dreaming isn't measurable. It isn't even rational. Its timelessness derives more from its illogic than from its pace. Nighttime dreams remove us from everyday reality, as our imagination does when we're awake, but the reality it offers is shallow, disjointed, and nonsensical. It can't be interaction with another reality.

On the other hand, a daydream allows our imagination to flourish without the irrational aspect. A daydream can provide the spark for human achievement. "Follow your dream!" is an expression that elders offer to youth. Wisdom teaches that if there's a cause for regret as we age, it's often the failure to pursue an idea that had germinated in our youth, possibly in a daydream, but had appeared unfeasible at the time. So what is a daydream? What makes it special? And how do daydreams relate to interaction with BTBB reality?

Daydreams, though more sensible than nighttime dreams, can be just as impractical. They often have no structure; they don't involve complicated planning or carefully thought-out procedures. Daydreams allow the imagination to blossom without restriction. Success can be guaranteed and insurmountable obstacles overcome with ease. We might laugh or cry if the daydream is strong enough, but usually we're merely suspended for a moment in a reality of choice, pondering what might be.

Like any other human activity, some individuals daydream to an unhealthy extreme, allowing the dreams to take over their lives. If the daydreams are so frequent that they affect everyday reality and inhibit the ability to function properly in that reality, they're destructive. So in searching for positive interaction with BTBB reality, we will disallow the aberrations of excessive daydreaming. Instead we will focus on the normal, fleeting thought process that pops up spontaneously, lasts only for a short time—although time can't be a factor if the BTBB interaction is legitimate—and leaves a feeling of euphoria when the daydream is positive, or a feeling of dread if negative.

A daydream is pure imagination, and, though it can be impractical, it isn't fraudulent. That is, a daydream is personal and therefore genuine to the person having it. The claim is fantasy and the daydreamer knows it.

The daydream may be based on our normal circumstance, or it may place us in an imaginary realm where we are the leader, the star, or the hero. It may involve a fantastic achievement in outer space or something as mundane as winning an award for redecorating the living room. Even if the subject is ordinary, however, there is a marked contrast between a daydream and actually planning a project, which, for redecorating, might include moving heavy furniture and protecting an expensive carpet. Real thoughts present real ATB problems. Daydreams don't.

Consider another example, a trip to the zoo. In ATB reality, plans must be made. First, there are the logistics: what route to take and when to arrive, since the porpoise show that the kids would love could be sold out early. Second, the trip requires choosing what to wear—it might rain. There are also decisions about food and drink, since past experience has taught us that the local zoo has a concession stand that charges a fortune, and the food is either too salty, too fatty, or too sugary. These are the normal, everyday thoughts required to get something done in ATB reality. They don't qualify for interaction with BTBB reality.

On the other hand, if we merely daydream about going to the zoo, we ignore all the ATB issues and think about other possibilities. We may imagine being fearless in the midst of lions or alligators, curing an aging elephant of a disabling disease, or training the porpoises to perform a new and exciting act. Daydreams ignore everyday problems and concentrate on the unobtainable. They wouldn't be daydreams otherwise.

A housewife dreams of romantic interludes. An actress dreams of stardom. A laborer dreams of making a fortune by inventing a clever new gadget. A clerk dreams of being on the other side of the counter and making the same unreasonable demands that the customers do. These are spontaneous dreams that quickly pass and don't interfere with the job at hand. The housewife wipes up the spill and goes on with her day. The actress reads the line she's read seventy times before, trying to rearrange her emphasis in a way that finally satisfies the demanding director. The laborer pushes the wheelbarrow full of stone to the walkway, dumps the load, and goes back for another. The clerk smiles and wishes the customer a good day. They all go back to the real world, the world that must function if they are to survive.

But what about that split-second when they pondered something better? What reality were they interacting with then?

Time and matter disappear at the critical moment of a daydream, thus providing the potential for interaction with BTBB reality. Of course, during the daydream ATB time marches on. If the daydream interferes with the job at hand, we quickly recognize our malfunction and apologize to any witness. Meanwhile, we've had an interesting experience that's well worth investigating. Let's explore what actually happened during the daydream.

What about the daydream's apex? There the idea of that perfect reality surfaces in a timeless manner then quickly disappears as ATB reality kicks back in. That timeless moment can be very precious. It may suggest the next step in mankind's progress. It may provide the germ of an idea that can lead to creating literary prose, discovering unknown territory, or finding a new direction for a business enterprise. It is a timeless moment that could be BTBB interaction. However, the practical aspect of the daydream may cloud the BTBB component. By definition, daydreams that help with everyday reality are more dependent on time and matter. Therefore, fanciful and impractical daydreams, which are more independent of time and matter, offer a better opportunity to examine interaction with BTBB reality.

Fantasy is a good example. A furtive glance at a pretty girl conjures up fantasy, even for a middle-aged, happily married man. Just appreciating beauty offers value. At that instant, time doesn't matter, nor does place. Thoughts aren't organized, and there's no plan to act on what the man sees. Yet, for that instant, the admirer remembers what once was, considers what might be, and acknowledges what **is,** all at the same time. Perhaps there's gratitude for the beauty of the opposite sex, perhaps there are joyous memories of youth, or perhaps there are thoughts of a young daughter's possibilities.

For the man glancing at a girl, the object of his fantasy isn't important, since the pretty girl is merely the tool he uses to slide into interaction with another reality. Once there, the mind experiences a deluge of thoughts that may or may not have anything to do with her. Instead, the mind takes a detour, even thinking of something as common as an interesting shape or texture. A powerful aroma or sweet sound that reminds a witness of a childhood memory may evoke a similar result. Aromas often stimulate a memory that sends the daydreamer into a different type of fantasy where the past becomes utopia. That "memory" fantasy may be a very effective tool for escaping from the present into a place where time and matter are unimportant. But whether it's the sight of a pretty girl or the aroma of freshly baked bread, the actual moment of escape is what's critical.

What happens at that moment? There may be the previously mentioned euphoria that, although quite temporary, is superior to most of the other experiences of our daily life. We may be flushed with satisfaction and accomplishment, even though we've accomplished nothing for our "real world" existence. Nevertheless, our egos inflate, our minds clear, and our hopes heighten. At that moment, there is no failure, no frustration, and no regret. Neither is there necessarily the desire to prolong the moment. Everyday reality must return, and we understand that. To live a normal life, we'll allow it to, but meanwhile the experience was ours and ours alone. A passing comment about the universal beauty of a pretty face might come from our lips or from a companion's lips as we come back to everyday reality, savoring the memory of a fleeting, indefinable moment.

So what do we have? What has the impractical daydream given us? It appears to be nothing other than a blurry memory. An impractical daydream appears no more helpful than drug-induced euphoria.

However, there **is** value in a daydream, and that conclusion becomes obvious by considering the person who doesn't have impractical daydreams, someone who can only live for the everyday reality that we're all immersed in. Think of choosing a spouse who never has a fantasy, who never pauses for a moment to come up with a thought that is unrelated to the issue at hand. What a dull person that would be. Few would wish to spend a lifetime with such a bore.

"A penny for your thoughts," is an old expression that demonstrates the curiosity of others in our daydreams. Without dreams life would be dull, and humans would be unimaginative, plodding creatures who never progressed. Although usually unproductive in itself, daydreaming allows us to see beyond our immediate horizon, to break rules and enter new spheres of reality that we might never have discovered otherwise. And even if these dreams don't affect our real life, we dream on, enjoying the reprieve that the daydream provides.

But there is more to impractical daydreams than just relief, even if the daydreams can never come true. They give us a sense of power over the forces that restrict our freedom. In daydreams we have no fear of a tyrannical boss or a demanding spouse. The experience may not be "real" as the rest of the world defines reality, but at the moment of the daydream, it's real to us. Even if the daydream doesn't lead to tangible improvement, it offers hope, and hope is a powerful tool.

Although we might like to change everyday reality into the reality of the daydream, the bubble always bursts and with it any possibility of the dream lasting. We're suddenly back in our everyday world, the one that we all agree is the **real** world. Just as we easily accept the "reality" of the daydream, we readily reject that reality when the daydream ends and true, gritty reality with its time and matter resurfaces. Meanwhile, we have had what we believe was a worthwhile experience, since for a fleeting moment it gave us hope.

Daydreaming is timeless. The actual instant of a daydream, such as imagined ownership of an attractive automobile, flashes in a person's mind. The clock doesn't matter. It only begins to matter when the daydreamer is snapped back to ATB reality. Daydreaming is nonmaterial and universal as well. It may be based on ATB reality, but physical laws are overcome while daydreaming. That's one factor that makes daydreaming so attractive. Also, daydreaming is universal. We all do it to some extent. Finally, there's no reason for daydreaming to exist in ATB reality. Any explanation of a daydream using ATB reality would have to be forced and unconvincing. Humans may like fantasy, but ATB reality can't explain why. Thus, daydreaming meets our criteria and gives us further understanding of what interaction with BTBB reality is and how it can affect our everyday life. And, as mentioned earlier, since daydreaming is personal, it has to be genuine unless we insist on deluding ourselves. As long as we admit it **is** a daydream, it's genuine.

Arguably everything we've discussed, from the various aspects of imagination to drawing a significant, abstract conclusion, could be considered normal human activity having nothing to do with a different timeframe. Nevertheless, we **do** have these experiences, and, at some level, they all meet the criteria we set up for interaction with BTBB reality. Since ATB interaction does not provide a satisfactory reason for having these experiences, there must be something more to reality to provide the explanation.

That "something more" is BTBB reality. Thus, an understanding of interaction with BTBB reality gives us another tool to use as a foundation for our understanding of life. With that tool we can make an objective analysis of the implications of human interaction and respond accordingly.

Before we put our newfound understanding to work, however, perhaps it would be beneficial to see if we can choose to interact with BTBB reality at any time we want. Has mankind ever tried that? Let's have a look.

Chapter 20
Generating Interaction

We have found interaction with BTBB reality in many aspects of the imagination, but it has all been of lower quality than the experience of reaching a significant, abstract conclusion. And face it, for most of us a significant, abstract discovery is unlikely. We're lucky to have one in a lifetime. So at this point, the average reader may feel that it's impossible to ever experience an intense BTBB interaction. There may be a way, however, to have one as often as we choose. To find such a method we'll go back to that challenging discipline—mathematics.

In math there is an occasion where everyday reality doesn't fit in with the numerical system. It happens when the answer to a problem includes the square root of negative one. The square root of a number is another number that, when multiplied by itself, gives the original one. For example, the square root of four is two, because $2 \times 2 = 4$. The square root of nine is three: $3 \times 3 = 9$. Some numbers don't have an exact square root, so the square root must be approximated by a fraction. The square root of 5 is 2.236068. If 2.236068 is multiplied by itself the answer is 5.0000001. Not exactly five but close enough for most applications. All numbers, therefore, have a square root of some sort, either an exact number or a fractional approximation.

Negative one, however, is an exception. It has no square root. The reason is that positive one times positive one is positive one. More important, negative one times negative one is also **positive** one. The only way to get negative one is to multiply positive one by negative one, and, since those numbers aren't identical, neither can be the square root. Thus, a negative number throws a monkey wrench into calculations involving square roots.

To solve this dilemma, mathematicians call the square root of negative one an *imaginary* number and assign it a symbol which they've labeled *i*. They then go ahead with their calculations, which proceed seamlessly, and the use of this imaginary number offers no impediment to solving problems. In fact, it's such a routine process that it wouldn't be worth mentioning except for one thing: the real world, the world of ATB reality, has actual mathematical solutions that have *i* in the answer. While most of us don't perform such calculations, we all use electricity, and, in problems about electricity, some solutions use *i*. Of course, most of us couldn't care less about equations involving electrical circuits, but that isn't the point.

To get to the point, consider the following. If I have two apples and my wife takes one away, I'll only have one left. That we understand. The number system works fine. But if I presented another situation where simple subtraction wouldn't work, and a convenient method to get the answer required the square root of a negative number—there are plenty of problems like that in electrical circuitry—you'd have to include *i* in the correct answer. That is, you'd have to use a completely cerebral term invented by mathematicians, one that logically shouldn't exist, to solve a real-world situation.

The argument could be made that the numbering system itself is cerebral, made up to reflect what we experience in everyday reality. As true as that is, however, that numbering system does not properly account for the square root of negative one. Everyday reality only forces the issue indirectly. That is, since there are some mathematical solutions that require the square root of negative one in the answer to a real-world issue, the real world forces mathematicians to invent a term to account for this aberration.

Perhaps the fact that it is indirect gives us the ability, when considering the square root of minus one, to send consciousness into interaction with BTBB reality.

Such a maneuver isn't as odd as you may think. For centuries people have been trying to find interaction with another reality. Zen meditation is one example where the practitioner attempts to put the mind into a state of suspension: a relaxed, perfectly conditioned state that involves neither time nor matter. Those meditating attempt to empty the mind of all thoughts by concentrating on the scent of a flower or the sound of a word—anything that's neutral. The purpose of the exercise is to escape from the reality that normally bombards the senses. If the effort is purely cerebral, it could launch one into a state of interaction with BTBB reality.

Perhaps considering *i* and trying to understand what *i* actually means could be a way of achieving a state similar to that of drawing a significant, abstract conclusion. It wouldn't be as exciting or fulfilling as reaching that conclusion, of course, and it certainly wouldn't be as effortless as daydreaming. Also, it's still just an approximation of interaction with BTBB reality. But even with those limitations, such an effort offers the possibility of additional insight into our understanding of BTBB interaction. We'll give it a try, but remember, in reading this you're using your eyes, and that keeps you in ATB reality. With that in mind, let's begin.

> What does negative one mean? It represents a quantity of one less than zero. If I have less than one apple, I would have to acquire one apple just to get to zero apples. It would be as if I owed my wife an apple, so if I got one and gave it to her, I wouldn't owe her one, but I wouldn't have the apple either. What do several minus ones mean? Say I have two of them. I owe both my wife and you an apple. That's minus two for me.

> Now let's try multiplying instead of adding. Multiplying is different. In multiplication a number is added together the number of times called for by the multiplier. So if the multiplier is two and the

number is six, two sixes are added to get twelve. If six is the multiplier, six twos are added and the answer is the also twelve. Minus one multiplied by minus one means that minus one must be added one less time than zero. Or, I must owe one apple one less time. If I remove the debt of one apple one less time, I'd still have the apple, and I wouldn't owe any. That means that minus one times minus one is **plus** one.

Next let's look at what a square root is. What's so fundamental about it? When does anyone use a square root?

Remember that triangle we used to demonstrate the relativity of time? In that case we used the equation, $A^2 + B^2 = C^2$. To get the value of C we must add the squares of both A and B and then take the square root of the answer. Here, all numbers are positive because squaring a number always gives a positive result.

Are we together? Are we interacting with BTBB reality yet? Or do you simply have a splitting headache and wish we'd move on? Be patient and try to follow along just a bit more.

There are times when the square root is used just as it was when demonstrating the relativity of time, but the number under the square root is negative. As we've already noted, some solutions to problems involving electrical circuitry are an example. It's not important what the conditions are, but it is important that the conditions arise. The question is why? Why would the square root of negative one come up at all? What is it actually? We know what plus one is. That's when I've still got that damned apple. We know what minus one is. That's when I owe you one. But what is the square root of minus one apple?

Think about it. Come up with an answer if you can. If you've been following this headache-inducing argument, trying to find that answer might only make the headache worse. On the other hand, you might be mesmerized by the problem to the extent that you find time suspended, and you begin to interact with another reality, the BTBB reality. It's worth a try.

People have long valued mental exercises, and many have found the process so uplifting that they've devised philosophical mind-stretching exercises to duplicate the experience. How many angels can dance on the head of a pin? Does a tree falling in a deserted forest make a sound? Over the ages mankind has posed these unanswerable quandaries. The disinterested may only find them somewhat humorous, but those involved find them passionately intriguing. To the involved, these questions send the mind into another state. As a personal choice, however, I prefer a mathematical concept, since it's an actual situation instead of a theoretical dilemma purposely set up to challenge the psyche.

The Square Root of Negative One.

$$\sqrt{-1}$$

Those of you who find mathematics tedious—I understand that includes a multitude—might prefer something else. A suitable alternative could be the Möbius Strip. A Möbius Strip is one dimensional and easy to make. Simply cut a piece of paper into a strip of whatever length you choose, twist the ends of the strip 180°, and tape the ends together. Take a pencil and trace a line along the middle of the strip, and you'll notice that you meet the point where you began without lifting the pencil off the paper. Thus the two-dimensional piece of paper is now one dimensional. Yet if you cut out any section of the strip, it has both a top and bottom. So what is the strip, really? Is it one dimensional or two? The complete strip appears to be one, but any cross section appears to be two. How can that be?

The Möbius Strip may not be as profound as the square root of negative one, and it's a physical rather than a cerebral example, but using it may work better for some people. Whatever works is fine. Perhaps counting those angels is better, or listening for that sound of the tree falling. The point is that if the contemplation is intense, the consciousness of time will disappear along with matter (even the paper the Möbius Strip was made from), and another reality will surface, one that is nonmaterial, timeless, and universal.

None of these conditions are perfect, however. Timelessness is violated in that there isn't the huge change in an infinitesimally small interval that our definition of timelessness requires. But if the quality of the meditation is high enough, it can approximate the interaction we're considering. Good meditation pushes everyday reality deep into the background and allows a new reality to surface.

Unfortunately ATB reality, though in the background, doesn't ever go away completely. Eventually it will impose itself, destroying the contemplative rapport that's been established and demanding something physical, such as nourishment. Nevertheless, for a few precious moments, a person may experience at least the hint of intense interaction with BTBB reality.

Mathematics frequently transports its experts into such a state by predicting enigmatic conclusions, like black holes. Often mathematical predictions that appear impossible at first glance, later on prove to be correct in observable, everyday reality. The math of String Theory predicts many more dimensions than the four (including time) that we find commonplace, and some scientists have attempted to find these extra dimensions. So although contemplating some ancient quandary, like the number of angels on a pinhead, doesn't help with everyday reality, pondering mathematical issues often does. It can reveal the way our everyday reality actually operates.

Since ancient times, mankind has been anxious to explain our ATB reality. Progress has been made, one step at a time, building on the discoveries of those who came before. Arguably much of the recent success has come from scientific experimentation and investigation using space probes and genetic manipulation. But there have been other significant discoveries which have originated in pure thought, from contemplating what we've observed and what the limits of our everyday reality are.

All that thought, all those conclusions, transcend time and matter and imply that there's far more to existence than the ATB reality that keeps demanding so much from us. Einstein's discovery wasn't considered part of everyday reality when he found it. But his work has led us to a far deeper understanding of reality. Now we can acknowledge that our everyday reality intersects with the BTBB reality. Think of the opportunity this presents. The BTBB reality is the one that existed before there was matter and before there was time. It's a real reality, not a made-up one that has no more meaning than those angels dancing. When we consider this other reality that existed Before The Big Bang and is still in existence, we're intersecting with another timeframe that's going much faster than ours. The billions and billions of years that we measure have an even longer duration in the BTBB timeframe. So when we proudly announce how long our sun will last, perhaps we shouldn't be so cocky that our clocks are the only ones that count. Our timeframe isn't the only timeframe. Even though it's the one we use as a reference, that doesn't rule out the existence of other timeframes out there.

If we could interact with BTBB reality in a pure form, we would understand this other point of view better. As it is, we have to work hard to approach such interaction, and when we do, we're limited by the quality of our experience. Nevertheless, in those moments, particularly when we concentrate on a very abstract idea, we can use the opportunity to discover interaction with a universe that isn't constrained by the physical laws that define our ATB existence.

Before the Big Bang there was no life, no universe, nothing that we associate with ATB reality. But there had to be some sort of reality BTBB, because if there wasn't, there wouldn't be any reality now. The Big Bang triggered both matter and time. That is, time began, our time, and, according to our measurement of time, life took billions of years to surface. We can theorize all we want about our present situation, explaining evolution and the universe in the terms dictated by our own timeframe, but eventually we must ask ourselves about the original reality that we came from.

*   *   *

Make no mistake. BTBB reality exists within ourselves, and we interact with it, albeit imperfectly, all the time. Understanding that interaction is important if we're ever going to understand our place in the universe. BTBB interaction is fundamental to our existence, since it's an interaction with a reality that is far more basic than the one we experience on the physical level. It came from before the beginning of time. It should help explain away what we label "The Mysteries of Life" and should be equally effective in helping us with ATB reality, since BTBB reality is the basis upon which our current reality is structured.

Our difficult search has ended. Now that we have identified forms of interaction with BTBB reality, as imperfect as they are, the question becomes: how does knowing about such interaction help us? How does BTBB reality influence us? And what is the next step? If we can interact with BTBB reality from our perspective, can another reality look back on us?

PART THREE

THE IMPLICATIONS OF INTERACTING

WITH BTBB REALITY

Chapter 21
The Issues of Genuineness

Armed with the knowledge that humans interact with BTBB reality, we can now discover revealing insights into human life. We'll learn why some of us are Doers and others Dreamers, and we'll explore the extremes of those inclinations, the Manic Doers and Pseudo Dreamers. In Chapter 23 we'll begin our investigation of how BTBB interaction affects human activity. Before we do that, however, we must take another look at an important consideration. Since all BTBB interactions are personal, to share others' experiences satisfactorily requires that we believe the interactions are genuine—that they're not a sham or part of a scheme meant to deceive or take advantage of a situation. In the course of this chapter, we'll note three issues relating to genuineness when considering BTBB reality and explore one of them in depth—the time and space difference.

Recall that in Chapter 15 we discussed the issue of genuineness in interacting with BTBB reality as it applied to the imagination. We recognized the difficulty in confirming genuineness and decided, therefore, to concentrate only on those aspects of the imagination that we could confirm as genuine, such as humor or the imagination of a child. Now we'll consider the issue of genuineness for all BTBB interaction and see if we can come up with criteria to assure ourselves that what others claim to be a genuine experience actually is.

This is not a trivial matter. We must weed out all illegitimate claims of BTBB interaction, so that any discoveries we make are based on sincere, genuine, human activity. For two chapters we'll be developing a method which will help us establish that the interaction we're considering is genuine BTBB interaction. If we glossed over this problem and went directly to our exploration of the effects of human interaction with BTBB reality, we would run the risk of making observations based on a false premise, nullifying any discoveries we made.

Remember, people have always had the ability to interact with BTBB reality, even though mankind has never understood what that interaction was. These interactions are no different today than they ever were, but, just as there have been disreputable individuals who have misused these experiences in the past, there will be others who will misuse BTBB interactions in the future, whether or not they choose to identify them as such. Thus, the ability to detect genuineness is critical and must be pursued, even though it presents a difficult challenge.

Since BTBB interaction occurs within an individual, confirming its genuineness is dependent on the individual's veracity. We've specified that true interaction must be observable and measurable, but, since the interaction is personal, observations and measurements may be significantly different from one person to the next, based on a variety of backgrounds and interests. Unfortunately that leaves a door wide open for chicanery.

When we are trying to determine whether another person's BTBB interaction is genuine or not, we have only two yardsticks to use in making our judgment—the resultant idea and its author. Notice that making that determination is significantly different from merely asking if an interaction is with BTBB reality. For that we have criteria, and, even though the criteria are so involved that it took much of this book to develop them, they are defined and workable. Contrast that with examining another's claim of BTBB interaction for genuineness, where, as we have stated, we can only examine the resultant idea and evaluate its source to make a decision. Accordingly, we'll only be able to make an estimate, since we cannot read another person's mind. Nevertheless, we must develop tools as best as we can, so that the estimate will be as accurate as possible. Please keep in mind that no matter how good our expertise is, our judgment of genuineness will only be that estimate.

This lack of certainty will be unsatisfying for some. Why even bring up genuineness if we can't be sure that our judgment is correct? The reason is that making an informed estimate is much better than either accepting or ignoring claims based on emotion or whim. Therefore, we'll establish tools for analyzing the genuineness of ideas that are the result of another's BTBB experience. We'll begin by identifying three issues that define that challenge.

Initially we note that these issues don't play a prominent role in cases where the interaction is primarily with ATB reality. That is, ATB interactions force objectivity. Hot things burn. Lemons taste tart. Jet engines roar. Sewage smells bad. Driving fast requires concentration.

Clear responses such as these, however, don't come with BTBB interaction. Each reader of a novel brings a unique background to its interpretation. Even for the highest quality of interaction with BTBB reality, such as what Einstein had when he recognized the relativity of time, the originator of the idea must work hard to convince others that the interaction is genuine and the explanation of the experience is correct. Convincing someone else of a revolutionary conclusion can be virtually impossible, unless the conclusion can be proven physically. Then, at least, we can agree that the interaction was interpreted correctly, and genuineness is more likely to be there. On the other hand, if the conclusion is proven false, then the interaction itself may have been genuine but misunderstood. Or it may just as easily have been insincere, misleading, and counterfeit. If a conclusion can't be demonstrated as accurate, it's simply interesting. Without a common, verifiable reference, any BTBB interaction may or may not be genuine. The difficulty is in finding that reference, especially if the interaction can't be easily related to a physical experience.

The first issue that affects the ability to detect genuineness, as noted earlier, is the fact that human interactions with BTBB reality are imperfect and intermixed with everyday, ATB reality interactions. Our daydreams and meditations occur while we are driving, walking, or sitting and enjoying a sunset. The second issue is that when we daydream or meditate or have any other form of BTBB interaction, our thoughts are unique to each of us, so attempts to compare our experience with another's can result in significant discrepancies. That is, we each experience a BTBB interaction in our own way. The exact method that BTBB reality enters into our experience varies. My aha moment may be quite different from yours both in time and content, even though we're both thinking about the same thing.

There's yet a third issue that surfaces when identifying the genuineness of an interaction with BTBB reality and which is, perhaps, the most challenging as well as the most helpful. This is the issue of time and space.

We've already discussed time at length, of course, but we haven't considered any spatial differences between the two realities. There is a good reason for that. Whereas the time difference is defined and measurable, the spatial difference is unknown and unworkable. It wouldn't be worth bringing up except that there **is** a difference in the space of the two realities, and any interaction with BTBB reality should have some spatial consequence. Unfortunately, since it is so nebulous, noting a spatial difference only offers a hint that another reality was encountered and nothing more. Nevertheless, both time **and** space are different when comparing ATB and BTBB reality, and it's important to recognize that fact.

Let's look at the time difference again. Recall that since BTBB reality is moving with respect to our ATB reality at virtually the speed of light, any interaction with BTBB reality is far different than an interaction with ATB reality. Think of how we struggled with the definition of timelessness in earlier chapters. The definition in itself proved challenging. A vast amount of change in a miniscule period of time is difficult to comprehend. Under the rules of ATB reality, such an experience would be chaotic. Knowing that all clocks run the same whether they're in ATB or

BTBB reality offers some consolation, but it doesn't resolve the issue. The vast difference in the speed of time surfaces when the two realities intersect, which occurs in any BTBB interaction that we have. It's the drastic change in the rate of time that provides the most startling and dramatic difference between interaction with ATB reality and interaction with BTBB reality. The purer the BTBB interaction, the more obvious this time differential is. In lower forms of BTBB interaction, however, timelessness is less perceptible.

While any timeframe difference presents a challenge in identifying genuineness, it can also be considered an aid, since it offers a measurable tool for identifying BTBB interaction. The more we understand that difference, the better we'll understand BTBB interaction, and the more we'll be able to judge the genuineness of another's interaction. So let's once again look at change and timelessness. This time, however, we'll concentrate on someone else's experience of change rather than our own. When we judge change through another person's experience, how is making that judgment different?

First, let's concede that most of us find change exciting and relish it. If anyone claims an impressive change in minimal time, the claim will impress the world. Humans love change. Think of the poor soul who loathes the dynamism of life and thus is missing out on much of what life has to offer. The stick-in-the-mud individual who's sorry the computer was ever invented not only fails to give proper credit for human advances where computers played a role but also misses the opportunities that computers offer. True, nostalgia has its place, and pining for the simplicity of the past has its value. But the reason is that going back to the past is in itself a change. Nostalgia confirms that change is exciting, even if the change is back to former ways.

So, although timelessness—tremendous change in an infinitesimally short time—is mind-boggling, it's tremendously exciting to the person experiencing it. In addition, when the experience can be adequately described in terms others can understand and accept, as Einstein was able to do in his publication on relativity, the experience becomes so precious that it stays with its author for a lifetime. Our challenge, therefore, is to recognize the timelessness of others when interacting with BTBB reality, so that it can help us distinguish genuine BTBB interactions, like Einstein's, from false ones.

That may be easier said than done, however. Time is the basis for our understanding of reality. When there's a dichotomy between two views of reality, and the dichotomy is dependent on the interpretation of time, reality appears distorted to both sides. So, if someone forms an opinion using interaction with BTBB reality as its basis, and someone else forms an opinion using interaction with ATB reality, the two opinions can differ to such an extent as to be irreconcilable. Only if both individuals temper their opinion with the understanding that there **are** two different timeframes to consider is a common understanding possible. Thus, the first step in judging the genuineness of any interaction is to accept and work with the two different timeframes.

Unfortunately, the practice of working with two timeframes is challenging, particularly when the timeframes are within the experience of someone else. In our case we're trying to use our expertise in comparing the two rates of time as an aid to determine if another's experience included a genuine interaction with BTBB reality. That complicates the issue even more.

When we compare our experience of timelessness with another's, we immediately run into the second issue, the individual differences of the experience. That is, no two individuals have the exact same interpretation of a BTBB interaction, since we can't interact with BTBB reality on a prefect basis. If we could, we'd have a single interpretation of that interaction with a common view of the amazing difference in the perception of time.

Again, imagine that we could have a perfect BTBB interaction. That is, think of looking at BTBB reality with our own ATB reality's clock. Remember, one second in ATB time would be eons of BTBB's time. Its equivalence in our timeframe would be as if World Wars I and II were intermixed with the chaos of the Middle Ages. Imagine witnessing a timeframe where the invention of the light bulb and the writing of the first hieroglyphics were seen in the same glance. All events would essentially be happening at once.

Yet horses and cars can't occupy the same streets concurrently. If time doesn't keep events apart, what else could?

That's where the spatial consideration kicks in. If time isn't the separator, space is. So, as the time difference helps identify the genuineness or legitimacy of BTBB interaction, the space difference should help as well. Arguably, of course, there was no space BTBB, but whether there was or not, the whole idea of spatial definition wasn't the same as ATB. In fact, comparing the spatial differences between ATB and BTBB reality, even though we know little about the latter, proves fascinating. Let's see why.

Mass curves space, and Before The Big Bang the mass of our universe didn't exist. Thus, there was no similar curvature of space. The Big Bang brought with it large masses that distort space. So not only did the Big Bang slow time, but it bent space, allowing gravity to keep us on earth.

BTBB reality has no mass, so its reality doesn't bend light. Its reality is linear. Yet we're so attuned to working with a reality where mass bends light, that, unless it's totally internal, operating in any interactive mode where no such distortion exists would certainly be difficult. If the interaction took place in our own thoughts, we could easily accept it as genuine, but if it took place in the thoughts of others, we'd be skeptical. When a great idea like Einstein's surfaces, it takes a significant amount of time for the idea to be understood. Eventually the idea's ramifications become clear and its justification is judged adequate, so most people will accept it. But that's not always the case. Remember, even Einstein himself challenged the probability aspect of quantum mechanics.

Therefore, anyone else's interaction with BTBB reality must appear different, not only because of the vast difference in the rate of time, but also because the interaction is with a reality where space is linear, since there is no mass to give it curvature. ATB reality has curved space and a defined rhythm of time. Straighten out space and interrupt that rhythm, and there is bound to be incomprehension. Yet though this incomprehension may appear to inhibit our search for genuineness, it actually helps, since it flags a different reality. Our challenge is to examine these differences of space and time and use them as a basis to make our judgment.

Another way to help us establish whether a BTBB interaction is genuine or not is to take a close look at ATB interactions for their validity. Many ATB interactions are truly genuine without question, e.g. hot irons burn. But other ATB interactions are not as self-evident. And—**this is what's exciting**—the ambiguities in these other interactions mirror the time and space differences of BTBB reality.

The best example of ambiguity in ATB reality can be found in our perception of ATB reality itself. Physicists have proved that this perception is often incorrect. Intuitively mankind has thought space to be linear and time to be absolute, but neither is true. If space were linear, we'd fall off the planet, and if time were truly absolute, the past wouldn't play the role it does in our existence.

To understand that misperception, let's look at the way the past plays a role in everyday human activity. Consider a basic ATB interaction, like eating. We know that if we eat too much, we'll gain weight, so we limit our intake of food. We realize, however, that eating one piece of candy isn't a problem. Perhaps we can eat two. Perhaps three. But we also understand that what we eat is cumulative, and our memory gives us the information to decide when to stop, even if we think we're still hungry. Our judgment is not based on one piece of candy but on the accumulation of how much we've eaten totally. It's based on a sequence of events, just as we intuitively think of everyday reality. Yet after we've made the judgment that we should stop eating, guilt intensifies with every extra piece of candy. In that way, the moment of guilt includes all the experience in the activity or *a lot of activity in one moment.* That's similar to what we defined as the timelessness of BTBB interaction. Thus, our ATB reality displays the timelessness of a BTBB interaction, and, just as important, this "timelessness" of the human memory is critical for survival.

People who have faulty memories can't survive on their own. Consider the Alzheimer's patient who can't remember the day of the week, the season of the year, or even what year it is. Eventually the patient doesn't recognize anything or anyone, even someone as close as a child. This memory loss would simply be unpleasant and inconvenient if only particular facts were forgotten. However, what makes the disease devastating is that the whole sequence of memory is affected. That is, since all of the past isn't included in a victim's thought process, an accurate judgment about life's choices can't be made. Without the ability to consider the complete picture, the Alzheimer's patient becomes lost and unable to function. Stoves will be left on; meals will be forgotten or eaten twice. No truly informed decisions can be made. Thus, memory is as important for humans to function as gravity is.

As we have seen, our intuition of linear space is incorrect when it considers ATB reality but correct for BTBB reality. In BTBB reality space **is** linear, since there is no mass to curve it. Why, then, would we intuitively think that space is linear and not curved when in everyday reality it **is** curved and in BTBB reality it isn't? In the same manner, our view of time is similar to the timelessness of BTBB interaction in the sense that our memories combine all of our past experiences with the present to allow us to make a decision about what needs to be done. We only function properly if we can remember a past that includes our previous experiences. So what we intuitively define as reality **is** reality for BTBB interaction but not for ATB interaction.

Thus, we've established that **both** ATB interactions and BTBB interactions have some ambiguity. In practice they both occur throughout the day. In the case of ATB interaction, the intuitive reality is incorrect, yet we use that intuitive reality as a standard to judge the validity of the interaction. In the case of BTBB reality, the intuitive reality is correct, but, since the interaction isn't physically self-evident, we consider the experience personal and open to question. Thus genuineness is an issue both for BTBB and ATB interactions.

To recognize genuine interaction with BTBB reality we've identified three issues: 1) that BTBB interactions are intertwined with ATB reality; 2) that BTBB interactions are unique to each individual, and, perhaps most important; 3) that two basic differences exist between BTBB and ATB reality—time and space. We've also discovered that what is intuitive for interaction with

ATB reality is actual for interaction with BTBB reality. Further, the timelessness that we would expect in interaction with BTBB reality actually allows us to function in the ATB world. That is, without considering the past as a whole, we would be unable to survive. In the next chapter we'll be developing tools to aid us in estimating the genuineness of BTBB interaction. The information above will prove helpful in defining those tools.

Finally, we note that the ambiguity of ATB interactions means that they aren't certain either. So when we ask what true reality actually is, we need to include the possibility that an ATB interaction isn't genuine. This makes our quest even more basic. We must define how we know whether any human interaction, ATB or BTBB, is genuine or fraudulent, and then we'll be able to judge the genuineness of BTBB interactions, just as we judge the genuineness of ATB ones.

Chapter 22
Tools for Judging Genuineness

How do we know if any human interaction, whether ATB **or** BTBB is genuine? For centuries philosophers and thinkers have raised questions about reality. Now, however, by recognizing that total reality includes both BTBB and ATB reality, we have new information to use as an aid in defining genuineness. By the end of this chapter, you will have help in making judgments about human interaction without having to rely on the so-called experts to do it for you.

To begin, let's consider an example of an ATB interaction. Touching a hot iron hurts, and it hurts a lot. There is no doubt in our minds that our skin is singed; it's an accepted fact. But is there anything more to this experience?

To answer that question, let's begin by considering the quality of ATB interactions in general. Are they all that pure? That is, arguments in previous chapters have insisted that BTBB interactions must have an ATB element. Perhaps a similar argument could be made for ATB interactions. True, we've discussed ATB interactions that appear to be only ATB—hot irons burn—but perhaps the burning isn't the complete interaction. There's also the scar from the burn, the suffering of the victim, and even the chastisement that more care should have been taken in the first place. In addition, although a swear word at the moment of contact reflects the immediate reaction of the victim, that word might also include the understanding of all of the consequences of the act: the pain, the necessity for treatment, the missing of an important appointment, and the like. The complete experience may include a timeless, nonmaterial, and universal element in its aftermath. So, while ATB reality is material and has the rhythm of our timeframe, there could be an argument that interaction with ATB reality includes some BTBB element when the whole experience is considered. Let's look for it in the hot-iron example.

Certainly there's no argument about whether hot irons burn. However, there could be aspects to an experience with a hot iron that are questionable. Perhaps the iron was supposed to be disconnected and cool. If there was a misunderstanding, whose fault was it? Perhaps there was a short in the switch, so that the iron was on when it should have been off. Perhaps the light that designated "ON" wasn't working, so you thought it was off. Of course, all of these conditions don't dispute the physical fact that hot things burn. But the whole experience includes both the immediate issue **and** the events surrounding it.

For this example, let's say the on/off switch wasn't working properly. Nevertheless, you could have noticed that the iron was hot as your finger got closer, but your mind was elsewhere. You had a problem at school and were thinking about how to handle it when you approached the iron. However, you find out that the reason the switch wasn't working was that your brother had dropped the iron on the floor. You also discover that he told your mother that when he checked the iron everything was fine. Then you learn that your brother lied; he knew all along that the switch was broken but didn't want to admit to it. When confronted, your brother claimed that he had intended to be honest but was worried that he'd miss his ballgame if he told the truth. He complained that he didn't have much time to consider the consequences, so he spoke on the spur of the moment.

Okay, were there any genuine Before The Big Bang interactions here? First, you know your brother would claim one, if he thought the claim would get him out of trouble. Second, you know your own mind was wandering. Thus, for yourself, you can easily assign your absentmindedness to a genuine BTBB interaction. It would be low-level, of course, but it was there. You were daydreaming. How about your brother? When he came up with the lie, was his quick thinking the result of BTBB interaction? It had timelessness, universality, and a nonmaterial element, so he could have had his own, very low-level BTBB interaction. On the other hand, your brother may just have been willing to say anything to keep out of trouble, and there was no genuine BTBB anything.

This example demonstrates that whether we're considering an interaction that's predominately with ATB **or** with BTBB reality, there can be a genuineness issue. Only the ATB component can be verified physically—that hot iron burned! The BTBB component may or may not have the capacity for physical verification. Einstein's Special Theory of Relativity was able to be verified experimentally, but other potential BTBB interactions, like that fib, can be impossible to physically prove. It would come down to your brother's word about his experience, and you don't trust your brother one bit. On the other hand, you know about the iron, and you can make your judgment based on what happened. You judge that the fib was just gibberish, fabricated to keep him out of trouble. If he had been more creative, you might have given greater weight to the possibility of his having had a BTBB interaction, but the iron story was too predictable.

Then again, debating the possibility of a BTBB interaction in so trivial a situation might be considered a waste of time. Perhaps it is, but the purpose of doing so is to make the case that, even in such a minor matter, judging whether the complete experience was genuine is difficult. Judging experiences which are more complicated and subtle presents even more problems. In such cases, the ATB aspect of the experience could be tougher to confirm, and the genuineness of the BTBB interaction might be even more questionable. The saving issue is that all genuine BTBB interactions have an ATB component, and that component can be verified physically. Even the most abstract philosophy relates somehow to the physical world that humans occupy. If it doesn't, people will pay no attention to it.

Repeating, no interaction can be purely BTBB. Thus, any claim of human interaction that doesn't have an ATB component must be false. That would mean pure interaction with BTBB reality with its timelessness providing an overview of an eternity's worth of time in an instant, an impossible human experience. Yet individuals make such claims frequently. Mind reading, fortunetelling, and other trickery are simple examples. Less obvious are claims of being able to cure disease or communicate with extraterrestrial beings, where the ATB component is implied but not obvious. People may claim "proof," but that proof is based on innuendo, unsubstantiated information, or even downright lies. Science itself can be mysterious for some, and that mystery invites such fraud. Perpetual motion machines have been "invented" over and over again.

To establish the genuineness of **any** human interaction, whether it be primarily ATB or BTBB, we can use the following criteria: **First,** as just noted, all human interactions must have an ATB component that's physically verifiable. That is, there must be a sensual observation that confirms the ATB aspect of the interaction. Once that is established, any genuine claim beyond the ATB component would have to be BTBB interaction. **Second,** for the BTBB component to be genuine, it must meet all the criteria defined for BTBB interactions—timelessness, nonmateriality, and universality. Further, this component cannot be explained by normal ATB interaction. **Third,** the interaction must be capable of being explained in a reasonable manner using accepted terms for normal, communicating adults. The explanation could draw from the ideas of Chapter 21, including the differences in the rates of time when the two realities intersect, the linearity of BTBB space, and the ways common adult perceptions of reality reflect interaction with BTBB reality. However, the explanation can't be mysterious in any sense of the word. It must use common logic and accepted forms of communication. Perhaps looking at examples would be the easiest way to see how these criteria work.

In the first case, we'll see what effect the verification of an ATB component can have on BTBB claims that were made using that component as a basis. This example considers a common ATB observation, the rising and setting of the sun. We observe it, and the observation is a genuine ATB interaction. The only way it wouldn't be genuine would be if sunrises and sunsets didn't occur but were invented to "explain" night and day, which we knew was actually caused by, say, an eclipse. If our senses sense something, and we sincerely report what they sense, then the observation—the report of the interaction—is genuine. However, we could report the same observation but incorrectly conclude that the sun was moving and the earth was still, rather than that the earth was rotating. That would be a genuine interaction with an incorrect conclusion.

When the conclusion is incorrect, there can be significant consequences of the error, even though the interaction was genuine. Assume that an observer uses the idea that the earth is stationary in order to evaluate the earth's significance within the universe. Thus, the sun would move and the earth would not, due to the fact that the earth is more important. This abstract conclusion, supposedly based on ATB reality, was the basis of many religious arguments in Pre-Copernican theories of the universe. However, the interpretation turned out to be based on an incorrect observation of ATB reality, and many other abstract conclusions reached using that interpretation were also proved to be incorrect. That is, the BTBB component of the ATB interaction, the abstract conclusion about the importance of the earth, was proved wrong by analyzing the ATB aspect of the interaction correctly. In doing so, common logic and accepted communication techniques were used.

With that understanding in mind, let's probe more deeply into using our criteria for establishing genuineness by applying them to an example that's primarily a BTBB interaction. Again, the criteria apply for **any** interaction. We're using a BTBB interaction this time instead of ATB reality as we did in the sunrise/sunset example.

Say a good friend makes a claim. She says that she's actually resolved the issue between quantum mechanics and general relativity and has successfully coupled the micro-universe with the macro-universe into one all-encompassing theory. Your reaction would be appropriately skeptical. Physicists have been working on this for decades, and your friend may be smart, but that smart?

The answer lies in the way we look at dimension, she claims. We look at dimension as being spatial—length, width, and height. Although we understand that time is a fourth dimension, since time is difficult to portray graphically, we fail to understand its full significance. She notes that previous attempts at resolving the incompatibility problem between the macro and micro-universe required the number of dimensions to grow in order for the theory to work. That's wrong, she insists. There is no such thing as dimension as defined in classical physics. Physical space is not only time dependent but continuous. Using length, width, and height to locate something is only a reasonable approximation that works for the macro-world. She then gives her own version of the continuous space she envisions. Her space isn't segmented as our macro-world space is. Distance depends on circumstance instead of physical separation.

You listen attentively. There's no doubt that your friend is convinced that she's made a significant discovery. Her abstract conclusion is exciting and earth-shattering, and it has all the earmarks of interaction with BTBB reality. If correct, it would be far more revolutionary than Einstein's discovery that time was relative. However, for you to accept her conclusion, you insist that she provide some kind of physical demonstration. If she could give that demonstration, you'd be convinced. Even if she couldn't verify her conclusion, however, if her reasoning was sufficiently adequate, as Einstein's was in his paper on relativity, you might become one of her supporters.

Notice the "ifs" in the previous paragraph. One required physical proof and another sufficient reasoning. Both of these "ifs" address the ATB aspect of the experience. However, even if both of these conditions are **not** met, you might still be drawn into your friend's thought process and attempt to develop it yourself, if your friend was convincing enough. And that's where genuineness comes in. Her argument must be strong enough for you to make a reasonable estimate about the genuineness of her interaction. Of course, it helps to know that she's a sincere person and not a scam-artist who enjoys making claims that can't be proved. Your friend is mature and reliable, so when she makes a claim, you know it's one she actually believes. However, when you note that her argument is sound and well-developed, it further strengthens her sincerity. In your eyes, her credibility is proven. You become a convert. Continuous space. What might it mean? Perhaps it's worth investigating. Since the idea is intriguing and offers tremendous potential, you team up with your friend. Perhaps successfully, perhaps not, but she has convinced you that she has had a genuine BTBB interaction. Has she?

Your answer is most likely yes, even if her conclusion is incorrect. She had a dramatic insight that was timeless, nonmaterial, and universal and which also had an ATB aspect. If her original insight was incorrect, it would follow that her conclusion would be wrong. The important point, however, is that in either case you would assume that she had had a genuine interaction with BTBB reality. Certainly her idea didn't come from an investigation of ATB reality alone. ATB reality includes three-dimensional space with time as a fourth dimension. Continuous space, where physical separation only plays a role due to defined conditions, is foreign to ATB reality. Your friend's spatial idea even touches on BTBB's linearity. Thus, only BTBB interaction provides a fully satisfactory explanation for her conclusion. She still has to convince others to agree with her, and the first step will be to prove the ATB component of her theory. But even if the ATB component turns out to be wrong, that outcome would not invalidate her interaction. It would simply mean that the logic concerning her interaction was wrong. She still had the interaction. It was predominately BTBB, and it was genuine. It blended BTBB reality with ATB reality which could then be used for verification. Finally, even though her idea was unconventional, her reasoning was solid and based on normal adult logic without any fantastic or paranormal elements that added mystery or intrigue.

A fair question is: if genuineness doesn't guarantee accuracy, what's the point in trying to make a judgment about it? The point is that unless the interaction is genuine, there's no sense in any discourse about the interaction at all. If there is no genuineness, there is nothing there. Any argument, conclusion, or logic that has a non-genuine interaction as its foundation is useless. So despite all the challenges that are involved in determining the genuineness of a BTBB interaction, the effort is well worth it, since it will help eliminate concepts that are built on falsehood.

With a judgment of genuineness, your friend's supporters could proceed, knowing that the idea will not crumble due to an insincere or made-up claim at its base. If her supporters couldn't make such a judgment, they could be taking a huge risk and wasting an enormous amount of effort on something that is meaningless.

Another question might be: why worry about whether an interaction is ATB or BTBB? The questioner would note that what's important is whether or not the interaction is valid and legitimate, not what we label it. The answer is that the two aspects of reality are very different, and they need to be evaluated in different ways. To ignore the difference or treat the two as they've been treated in the past can lead to wrong judgments and misleading discourse. Think of all the Dreamers who've been condemned and the thinkers who've been ignored, simply because their audience wasn't prepared for their jump in logic. Recognizing a BTBB interaction is important, because it reduces the difficulty in making a judgment about genuineness.

Let's try a third example. Say some author claims that humans not only interact with the everyday reality that all creatures experience but with **another** reality as well, the BTBB reality. If this bizarre proposal blends ATB reality with BTBB reality and takes into account the differences in time and space that the two realities present, it may have come from a genuine interaction with BTBB reality. If the author claims all of humanity has the capability of BTBB interaction, such that it's universal, then the author meets the universality criterion for genuine BTBB interaction. The abstract insight is nonmaterial and timeless, so those aspects of a genuine interaction are met as well. And if the author uses normal, well-accepted logic in the argument, it satisfies the third criterion. Obviously, this example uses our premise as its base, and, since the author's veracity hasn't been confirmed, the reader may still be unsatisfied. Nevertheless, it's an example of genuine BTBB interaction.

(A personal note:

When I began to write this manuscript, my intention was to demonstrate the timelessness that can occur when two timeframes interact. BTBB reality didn't come into my thoughts until it became apparent that there could only be two timeframes that are meaningful to our reality, the ATB timeframe and the one that came before it. The moment of this "aha event" did, in fact, appear timeless. Let me assure you that this discovery required no dramatic insight or great thought. Anyone could have had the same idea once they concentrated on the possibility of two timeframes interacting.)

Now let's consider examples of interaction where BTBB legitimacy is far more questionable. For instance, some claims are definitely illegitimate, such as those made by people who predict the future or communicate with the dead. Again, the universality of interaction with BTBB reality isn't there. Those who make these claims maintain that they have "special" power that other people don't possess. Thus, any "blend" of ATB reality is with a BTBB reality that isn't universal. These claims are generally easy to confirm as false.

Other claims, such as those asserting new insights into human activity, may be just as suspect but are more ambiguous, and therefore more difficult to disprove. Writers or advisors who claim a revolutionary capability to predict activity in areas such as the financial markets fall into this category. Consider a TV personality who says he has a foolproof scheme for making money by trading stocks. He offers complicated mathematical formulas and vivid graphical data to convince the audience of the scheme's legitimacy. What is the best way to respond?

First, if the claim isn't grounded on proven ATB reality, and if the mathematics appear too complicated and/or graphic displays too obscure, the challenge is immense. Even if the figures correspond to historical data, they won't necessarily work in the future. The first step in estimating genuineness is to firmly establish a blend with ATB reality. At this point one can usually decide whether to proceed or not, based on the results. Once the logic is proven correct, the next step would be to examine the idea for evidence of timelessness, nonmateriality, and universality. If all that can be established, there is the test to see if the scheme is explained using normal, accepted methods. If any part of the scheme can't be, it most likely isn't genuine.

Most people would decide that all that effort for a claim heard on television wouldn't be worthwhile. On the other hand, if the author of such a claim was a personal friend, as was the physicist, and you knew how sincere your friend was, you would be more inclined to make the effort. But even then, you would want your friend to provide sufficient information to convince you, as your physicist friend did in her case.

This discussion has defined the challenges associated with sharing genuine interactions with BTBB reality. We've learned that, with care, we can make reasonable estimates of genuineness. Fortunately the key is that all interactions have an ATB component which can be confirmed by human observation. If any interaction can't be verified, at least with regard to its ATB component, people won't agree on its authenticity.

Humans are social beings. We need to share. That's why we communicate our experiences, whether we have a willing audience or not. We write books, tell stories, argue, persuade, and preach, all in the attempt to share our interactions, whether they primarily involve ATB reality, or if they include significant interaction with BTBB reality. Those of the latter variety are the ones with the more skeptical audience, of course, because they're the ones more difficult to confirm as genuine and correct.

Total reality is the sum of ATB and BTBB reality. As in everything else in life, we must differentiate between a genuine quest for identifying reality and an illegitimate one. In the latter, the unscrupulous can take advantage of the difficulty in defining that reality as leverage for gaining influence and even domination over others. Illegitimate quests must be exposed for what they are. Since the potential for fraud is so great, the need for verifying genuineness is significant. Once we've made our best assessment of the genuineness of an interaction, we can test it for accuracy, so that we can share it with others with the confidence that we're sharing truth.

We've given examples of both genuine and non-genuine claims of BTBB interaction. Since some ATB interactions have a BTBB component to them, we can use what we've discovered about BTBB genuineness in analyzing ATB interactions as well. For the ATB component we prove genuineness by verification of the physical consequences. For the BTBB component we use our criteria as tools. In summary, they are:

1. BTBB interaction must include a blend of ATB reality, with its curved space and ATB timeframe, intermixed with BTBB reality's linear space and BTBB timeframe. (We've only touched on linear space in one example, and that most likely didn't help much. As already noted, the spatial difference between ATB and BTBB reality isn't as defined as the time differential, and so it is much more difficult to consider. Nevertheless, we must recognize the property and accept it when it appears.)

2. BTBB interaction must have the characteristics of timelessness, nonmateriality, and universality. Further, the experience cannot be adequately accounted for by normal ATB interaction.

3. BTBB interaction must be explainable using normal, acceptable logic drawing on everyday human experience.

It may not be easy to fully comprehend these tools on an abstract level. Fortunately they are better understood using examples, and they will become more and more obvious as we consider BTBB interaction and its implications. We will use them to judge if an interaction is genuine, and we will see how such interaction affects human motivation and activity.

BTBB reality is the reality we inherited from the Big Bang, and it is built into our psyches. BTBB reality is universal. There's nothing mysterious or incomprehensible about it, even though its characteristics are quite different from those used when interacting with ATB reality. Timelessness isn't easy to grasp, and, since BTBB reality is nonmaterial, all interaction with it is personal. But it's not supernatural or mysterious. It's **natural!**

Chapter 23
Dreamers and Doers

We've seen that people continually interact with total reality, which includes both ATB and BTBB reality. We can utilize this information in judging our own actions, as well as in evaluating the actions of others. By looking at human inclinations and responses, we'll see that both realities play a critical role in life, but they each do so in their own way. Let's examine this basic division of the two realities and see how it is fundamental for human activity.

First, ATB reality. The physical world demands respect. If society chooses to ignore the physical world, it will cease to exist, period. Bridges with faulty designs collapse; tornadoes destroy homes; and too much rain floods the land. To combat these catastrophes society has built in checks and balances that are designed to insure against incompetence and natural disasters. Insurance companies sell polices and invest premiums, so there will be sufficient funds available when misfortune happens. Meanwhile, governments build levies, upgrade building codes, and install warning alarms to minimize death and destruction. Nature continually reminds us that physical reality is critical to our survival. If we ignore the physical, we pay the consequences.

Even a mental mistake can cause a problem. Bad subtraction in a checkbook can lead to an overdraft. Forgetting an umbrella risks the danger of getting soaked. Experience has taught us that we must use all of our talent and ingenuity to keep ourselves safe.

There can be social penalties for ignoring physical issues as well. An unmowed lawn brings frowns from the neighbors. An unmended fence, unkempt home, or poor hygiene results in a bad reputation and its consequences. Thus, we keep our homes presentable and ourselves clean, not only for physical survival but for social acceptance.

Now let's consider BTBB reality. That too is pervasive, but, since it's less obvious, it's harder to pin down. The most noticeable difference between ATB and BTBB reality is the significant increase in the speed of time, but even that is so subtle that we think of it as merely an aberration or a mind game. So we continue to live our lives, unconsciously blending the BTBB timeframe into them. As a result, although BTBB reality is just as real and vital to human life, we can't analyze it the same way that we analyze ATB reality. So we'll approach BTBB reality differently. Fortunately there are people who prefer to concentrate on this other form of reality. An analysis of those who hold this point of view will provide an understanding of the BTBB component of reality. By comparing this segment of humanity to those who concentrate on ATB reality, we will gain a clearer view of BTBB reality. Who are these individuals and where do we find them?

It's not a difficult search. Simple observation of human behavior tells us that while most people agree that human activity extends beyond the physical, there are some who consider all physical demands to be secondary. People who concentrate on BTBB reality prefer to ignore the practical, turn a blind eye to the functional, and concentrate on what they regard as a superior aspect of life. Individuals with this preference place a high value on mental achievement, abstract thought, and art—anything intellectual. They pay as little attention to material needs as possible without risking their ability to survive. They do manage to survive one way or another, but it is often at considerable risk to themselves and the rest of society. Yet they pursue their version of reality with every bit as much passion as those who concentrate on the physical. For our purposes, let's give a label to those who advocate subordinating the physical and concentrating on the abstract—we'll call them Dreamers. But before we delve into the world of Dreamers, let's distinguish them from their counterparts, the Doers.

Those who concentrate on ATB reality, the Doers, place their emphasis on the physical. That is, these individuals give top priority to material achievement. They are the athletes, the entrepreneurs, the bankers, and the politicians. They consider accomplishment as anything that leads to what they define as a better life, and that definition includes physical comfort, sensual pleasure, and social status.

All people blend ATB and BTBB reality to some extent, of course. But the point here is that Doers place their main priority on ATB reality, and Dreamers place theirs on BTBB reality. Thus, there is a range of human activity, from those concentrating on the practical—The Doers—to those who concentrate on the fanciful— The Dreamers. Exactly where a person fits within that range depends on individual preference and the degree of commitment. Some people are quite balanced, addressing their physical needs adequately but also leaving plenty of room for the arts, religion, theater, contemplation, and abstract reasoning. Others, however, are extreme. We'll see that the personality types range from Manic Doers to Pseudo Dreamers, but the full spectrum won't be evident until we examine both types in more depth. For now we'll focus on a comparison of the Doers and the Dreamers.

Doers thrive; they have to. They provide the services and sustenance that humans require for existence. Doers push interaction with BTBB reality into the background. Why concentrate on anything that has no apparent impact on life? If a Doer can't perceive something with one of the five senses, if something is so intangible that there's a debate about its very existence, the Doer has little time for it. Doers are focused on the physical and the time-dependent reality that our bodies use. The more a Doer can interact with ATB reality, the more a Doer feels fulfilled. Doers can read others, anticipate needs, and manipulate clients, so that their own goals are met even if those goals conflict with those of their customers. The most accomplished Doers are wealthy, sexually successful with multiple partners, and have both the fame and status to complement their worldly conquests. They understand ATB reality and use it to the hilt. They bask in its comforts and become society's heroes.

Now for the Dreamers. They manage to survive as well, but often just barely. Those that are too insistent on ignoring the practical may get by relying on the efforts of others, whether through charity or government grants. If their dreams are strong enough, and if they have enough talent to document them, they may persuade others to buy that documentation, which could be in the form of a novel, a song, a statue—or even a controversial book proclaiming the existence of interaction with BTBB reality. Many Dreamers, however, lack sufficient talent to support themselves with their dream and must concede to the necessities of life, waiting tables or sweeping floors, all the while keeping the dream as their top priority.

Dreamers may be as sexually active as Doers, but not for the same reason. Dreamers aren't interested in conquest. They use sex as an aid to discover the reality that they crave. Rather than triumph, they search for ecstasy.

Dreamers' personal budgets often restrain their social opportunities, however. If a Dreamer eats out, it's either as a guest or at a low-budget eatery. If a Dreamer drives, the car is among the oldest and tiniest on the road. If a Dreamer owns a home, it's in a less desirable section of the community, and the house most likely needs paint. Dreamers consider such circumstances merely an inconvenience, a small price to pay for the freedom to dream. They may be every bit as satisfied as the bank president, even though that Doer spends Januarys in Florida and Julys touring France. A Dreamer's imagination compensates for any physical hardship and provides the satisfaction that others receive from material possessions. Dreamers—many of whom choose the academic or artistic world for a career—have a clear view of themselves and share that view spontaneously with anyone who will listen. They're often much more at ease with themselves than ambitious Doers are.

For the purposes of our discussion we've called a person who is dominantly practical, a Doer, and a person who is dominantly impractical, a Dreamer. Most of us lie somewhere along the spectrum from extreme Doer to extreme Dreamer, mixing in our dreams with the practical necessities of life and working out a reasonable compromise that allows us to satisfy both appetites. **The important point of this identification is that the two views of reality correspond to the two realities that are the topic of this discussion, ATB for Doers and BTBB for Dreamers.**

This division isn't forced. We all know individuals who place a different value on certain aspects of life than we do. Some spend all their free time on impractical hobbies that mean much more to them than their "everyday" occupation. Others are so embedded in the physical reality that they show no regard for the artistic or the whimsical. The division exists, and it's no coincidence that it conforms to the two realities that we've identified.

Reasonable individuals interact with both realities, however. Note that since physical reality insists that we respond to it, and, since no one can avoid the physical, most individuals are Doers. That is, most of us predominantly interact with ATB reality. Nevertheless Doers often find time to consider the impractical as well. They'll pursue the fine arts, for instance, because even the most hardened entrepreneur understands the need for "something more." On the other hand, BTBB reality is the one that's fundamental to everything else. Perhaps that's the reason that those who are inclined to interact predominantly with it are so passionate. Dreamers love their dreams, and they'll sacrifice a great deal for them. But again, both the Dreamer and the Doer, if they are normal, take both realities into account.

Although there are Doers who admire and even envy Dreamers, most people don't understand the Dreamer lifestyle. The reason may be that Dreamers tend to live on the edges of society, and that makes people unsure about them. Is the Dreamer sincere, or simply a bum who doesn't like to work? The fact that Dreamers have their own set of rules gnaws at Doers. Thus, though they may not be an enemy—they may even be a close friend or relative—they aren't the full-fledged members of society that most of us strive to be. Most Doers aren't impressed with Dreamers unless they have some tangible proof, such as a book or a work of art, that the Dreamer has something of value to offer.

Of course, Dreamers don't have much regard for Doers either. Dreamers consider ATB reality as inferior to a more fundamental one, which we've identified as BTBB reality. Dreamers don't use that term and probably wouldn't accept it, because BTBB reality is too concrete, too defined, even though it's as different from ATB reality as any reality can be.

In spite of this distrust and ill feeling, both of these personality types have their allies. Dreamers especially have a set of fans who envy the freedom that accompanies their ability to pursue whatever they wish, rather than conform to a code of conduct that earning a living requires. The uncle who can't hold a job but always has the gleam of discovery in his expression draws the curious ears of nieces and nephews. And there's the woman who lives by herself and spends most of her time on the porch watching others pass by. Although seemingly lonely, her expression is forever calm and reflective. Anyone noticing her stops and says hello. Dreamers have that magnetism, and thus, even though we may be suspicious of their sincerity, we often find them intriguing and exciting

All societies have their share of Dreamers who range from the bum to the genius. In the twentieth century, Western culture labeled some of the idlers as "Beatniks" and "Hippies," and many observers considered them with bemusement and humor. Unfortunately both of these groups had a reputation for using mind-bending drugs. Drug use is an unfortunate aberration, but even Dreamers who don't have a problem with drug addiction can have societal issues. Few Dreamers get adequate respect or recognition from society unless and until their genuineness is verified. But when it is, those who were regarded with suspicion and even contempt are suddenly deemed brilliant. Thus, individuals like Vincent Van Gogh, Albert Einstein, Jesus Christ, and many others are initially saddled with the reputation of being a Dreamer and receive little appreciation. Once society makes its discovery, however, it frequently rewards a Dreamer with societal honors. Note that this only takes place after the Dreamer's efforts are documented in a convincing manner, not before. Those who gain recognition in their lifetimes are the lucky ones.

Arguably Dreamers are more committed to their outlook on reality than Doers are. In order to live with themselves, they have to be. Their attitude forces them to live without much of what society offers. The Dreamer's commitment itself may be the most admirable quality that people recognize, allowing them not only to tolerate Dreamers, but even to support them through grants and gifts. Society sometimes makes a judgment that to be so committed and giving up so much must mean that the Dreamer's activity is genuine.

Even though it may play a significant role in gaining support, eventual recognition doesn't depend on the Dreamer's commitment alone. The same applies to a Doer. Commitment in itself does not assure success for either viewpoint. A Dreamer who accomplishes nothing and is never recognized may be far more dedicated than a moderately successful Doer, who just goes through the motions that a job requires. And Doers may fare no better. There have been unsuccessful Doers committed to curing a dreaded disease, but who never did; or to discovering extraterrestrial life, but couldn't; or to winning a sports championship, but failed. If they don't succeed in their effort, Doers can end up in the same dire straights as unrecognized Dreamers. In such cases, the Doer may be just as committed as a Dreamer and just as frustrated. The only difference is that the lack of material reward can be harder on the Doer than the lack of acknowledgement is for the Dreamer.

BTBB reality is vague. What drives a Dreamer to be so committed to it? Why are some individuals so dedicated to an idea that they're willing to deprive themselves of physical comforts just for an opportunity to work on it? What motivates anyone to ignore ATB reality and concentrate on some other reality that's so subtle that many question its very existence?

The answer lies in the Dreamer's ability to interact with BTBB reality and the rewards that interaction provides. Dreamers become experts at connecting with the abstract and the theoretical. They consider the insight they achieve as ample feedback for any lack of material gain. They are the ones who understand BTBB reality, and, if we want to obtain as clear a definition of that reality as possible, we need to learn to view reality as Dreamers do.

Doers and Dreamers are the two classes of people that make up all societies. They're the practical and the impractical, and, as already noted, they don't particularly like one another. The argument between Doers and Dreamers is familiar, and one we can use when interpreting interaction with BTBB reality. In this chapter we highlighted the differences in outlook between the two types of individuals. Since, to some extent, we're all familiar with the Doer/Dreamer mentality, we can use that familiarity in any analysis of human interaction. The more we understand the Doer/Dreamer mentality, the more proficient we will be at identifying an individual's outlook on reality. Once we have that proficiency, we can proceed to consider the BTBB reality component of that outlook and make a judgment on the sincerity and genuineness of that person's BTBB interactions.

Just thinking of Dreamers, the artists, authors, teachers, and thinkers that we know, helps us comprehend BTBB reality. Further, since we understand how Dreamers view life, we can often spot the phonies and can use that ability to judge BTBB interaction in general. That is, once we establish the genuineness of a Dreamer, we will have help in establishing the genuineness of that individual's interaction with BTBB reality. Confirming a Dreamer's genuineness isn't easy, of course, since it requires overcoming the same hurdles that were mentioned in the previous chapter. The difference is that here we're on more familiar turf. So let's delve a little more deeply into the Doer/Dreamer views of life.

Chapter 24
The Problems of Imbalance

When considering human interactions, we have already established the need for genuineness. To review, genuineness means that ATB interactions must be physically verifiable, and BTBB interactions must meet the criteria developed in Chapter 22. BTBB interactions must be timeless, nonmaterial, and universal. There should be a blend of ATB reality within the BTBB interaction, and the BTBB aspect has to be explainable using normal logic and natural human experience. Now we will add one other measure to assure a healthy attitude toward human interaction—balance. That is, there must be a **balance** between BTBB and ATB interactions.

One of the traits of wisdom may be the realization that we never discover anything that is actually new. Instead, we gain a clearer understanding of what has always been the case. Human pursuit of the concrete and the abstract, which parallel ATB and BTBB reality, is not a new idea. However, we now understand the reason behind that division in outlook. In this chapter we will further our analysis of interaction with the two realities by identifying the problems associated with an abnormal concentration on either one or the other. We will learn how difficult it can be to find a proper balance between ATB and BTBB reality. In the following chapters we will see how this difficulty can lead to troubled human behavior. Once we understand the causes and effects of the imbalance, we will be able to return to the subject and discover a way to find a balance that will work for each of us. Thus, although we will not solve the imbalance issue until later, its recognition will help us analyze human behavior.

Most people appreciate both aspects of reality and work with both. Still, there's a natural tendency to pursue the aspect of reality which appeals to us more, whether it be ATB or BTBB reality. That tendency can lead to problems if it is taken to an extreme.

Individuals have a tendency to condemn others, simply because they don't share the same priority when it comes to the basic division of reality. Thus, a Doer will criticize a Dreamer and vice versa. The accused will respond with a defense of his or her choice. That too can lead to an extreme position toward one of the realities, since the defense may exaggerate the chosen reality's advantages.

Just using the labels may be insulting to some, but interacting with BTBB reality is an actuality, make no mistake about it, and labeling any activity as such neither limits its effectiveness nor pigeonholes its possibilities. That is, assigning a Dreamer label to religion doesn't disparage it any more than assigning a Doer label to an ambitious enterprise does. The crux isn't being a Dreamer or a Doer; it's the degree of emphasis. That is, problems develop when the emphasis is placed too heavily on either one reality or the other.

Let's investigate an instance where people place their emphasis exclusively on one reality. These individuals may insist they know what's best, but the imbalance in their outlook can lead to serious difficulties for themselves, as well as those around them. The problem escalates when those involved are leaders, and their attempts to force society in one direction or the other can affect entire nations.

First, we'll examine individuals with an excessive concentration on ATB reality. We'll begin by looking at a successful businessperson, such as a car dealer, in a small community. That dealer might have a reputation for paying low wages, demanding long hours, and using hard-sell marketing. Nevertheless, if the locals want the brand of auto for sale and the price is low enough, that negative reputation won't discourage most buyers. To rationalizing minds, rumors of rip-offs and complaints from employees are only one side of the story. Potential customers may conclude that if the dealership was so disreputable, it wouldn't survive. So customers make their purchases and congratulate themselves on negotiating a good deal. The proprietor ends up with the sale, grabs the profit, and takes another step toward the ultimate goal of many Doers— wealth. Meanwhile, both the workers and the community suffer. Eventually, the effects of cutthroat business practices hurt too many people, and the whole community is weakened.

As the scope of a leader's influence broadens, the consequences reach much further. A national Doer leader insists on material advancement and physical accomplishment, and such a Doer-led society will focus on those goals. The result is a practical culture, favoring aggressive behavior. Walking over people to get whatever an individual wants is often seen as a necessary trade-off for success. The heroes of the society are heads of corporations, talented surgeons, sports heroes, and landowners. Dreamers in such a society are often labeled as curiosities or worse.

By definition, Doer leaders emphasize interaction with ATB reality, but if they are to be reasonably effective, they will allow for BTBB interaction as well. However, if the leader is fanatical, that is, insists on **only** ATB interaction and condemns all BTBB interaction, the results can be catastrophic. These leaders tend to consider other members of society as their pawns and use them to acquire material gain and power. They may even be willing to engage in aggressive warfare and massive violence to achieve their goals. While some may realize that there's a problem, if they have no resources to combat the leadership, they will be helpless to act.

At the other end of the spectrum are leaders who are committed exclusively to interaction with BTBB reality. On the local level, these may be religious fanatics or cult leaders whose single-mindedness can lead to anguish and pain among their followers and their families. If a movement expands to a wider scale, the leader can become just as destructive as a Doer fanatic. Perhaps Europe in the Middle Ages provides a good example of obsessive BTBB emphasis. Humanity at that time surely had its share of Doers, but the leaning toward interaction with BTBB reality was so strong that human progress was stymied until the fifteenth century. For this reason, some historians have labeled the era between the fall of the Roman Empire and the Renaissance as the Dark Ages. Religious leaders of the time preached that the primary purpose of human existence was to worship God and that sacrifice and suffering were a positive good in preparation for the hereafter. Therefore, they repressed efforts to improve present circumstances.

This is not to say that the Doers of that age were stamped out. The Crusades, for instance, while claiming religious inspiration, were often aggressive Doer attempts to achieve power and wealth. The Doer Crusaders used religion for their own purposes in a Dreamer society and managed to leverage Dreamer goals for their own ends. This example shows that a Doer in a Dreamer society can more easily succeed than a Dreamer in a Doer society. ATB reality is too immediate to stifle its requirements completely, and no society can fully suppress the ambition of aggressive Doers.

The culture of the Middle Ages faded, because eventually mankind had to acknowledge that ATB reality existed and needed to be addressed. Subordinating ATB reality to BTBB reality will only work if ATB reality is adequately taken into account. Any society that is controlled by fanatical Dreamers who place total emphasis on interaction with the spiritual is doomed.

Great religious movements do not behave in this manner. These movements are effective because they focus on **both** BTBB and ATB reality. Examples are the Mormon exodus from Illinois, the Christian monastic movement of St. Benedict, and Islamic society during the Middle Ages. A shallow judgment might conclude that these religious societies overemphasized interaction with BTBB reality, and that they used ATB reality only to the extent that it was required for survival. A deeper look, however, shows a surprising balance of both. The impressive Mormon journey from Nauvoo to Salt Lake City wouldn't have been successful without a concentrated effort on every detail necessary to make the journey. The long trek required logistics that would challenge twenty-first-century social planners with computerized tools at their disposal. St. Benedict's rule carefully balanced prayer and work and gave both equal footing, claiming that work **is** prayer. Islam, especially from the eighth to the twelfth centuries, valued intellectual insight that had widespread practical application. Mathematics, architecture, and medicine thrived in that society.

So it would be incorrect to conclude that only fanatic Dreamers lead major religious endeavors. Religion itself is more than interaction with BTBB reality. To legitimate believers, religion encompasses everything, including BTBB **and** ATB interactions. In fact, religion is full of symbolism and physical rituals based on ATB interaction. Even prayer is often a request for improvement in an aspect of ATB reality. If either reality is emphasized such that the other is considered trivial, priorities get perverted, total reality gets warped, and failure becomes inevitable. St. Benedict understood that, and so did Brigham Young. Prominent leaders, both religious and secular, recognize this fact and administer accordingly. They don't identify the two realities as ATB and BTBB, of course, but their priorities demonstrate the equivalent.

Showing that balance between the two realities is necessary leads us to the question of what the proper balance is and how that balance is to be found. For instance, is it proper to condone a work of art that desecrates the flag? Doers would most likely argue that it is not, since the flag is a physical representation of what a nation stands for (even though the nation's values may include the freedom to practice such art). Most Dreamers, however, would say yes. Since the flag is only a symbol of a society that isn't perfect, what's wrong with desecrating its flag to make a point?

Doers and Dreamers have different opinions about what constitutes a proper balance, that is, how much of the other reality they should allow into their lives. Both are usually quite certain of their viewpoint and aren't easily convinced to modify it based on some opposing philosophy. Dreamers have an embedded opinion that their cause is the "higher" reality, the "greater" good. Anyone who's had a creative thought realizes the tremendous satisfaction and euphoria that accompany such an idea. To a Dreamer that experience is incomparable to, say, winning a massive lottery prize. But to a Doer the massive lottery prize offers much more, unless, of course, the Doer thinks more money can be earned from the creative thought than won in the lottery. Those Doers still wouldn't admit that interaction with BTBB reality is just as important as with ATB reality.

Perhaps a good balance would be easier to recognize when looking within ourselves, since we don't need an exterior method of communication for self-examination. That is, art, religion, and other BTBB interactions require some medium in which to exchange ideas with others, and that communication must be physical. However, if we interact with BTBB reality internally, through meditation for instance, no physical communication is required.

In meditation we contemplate all reality. We can meditate about the concrete or the abstract, but either way, as long as we're just meditating, we're interacting more with BTBB than ATB reality. If the meditation doesn't interfere with life, then the meditation can be positive, especially if it adds to our understanding of ourselves and others. On the other hand, if the meditation interferes with life, if, for example, we don't eat properly because of it, then the meditation is negative. Even though it may add to our understanding, the cost of that understanding is our physical health.

Unfortunately, in actual life experience, the results are often too ambiguous to make a clear-cut judgment. We may be meditating so much that we're not eating as healthily as we should, but we do satisfy our hunger. Our home may not be as comfortable as it could be, but it's a roof over our head, so we keep meditating. The balance may be adequate in our opinion, but inadequate as far as others are concerned. The judgment is subjective. The meditator believes there is sufficient balance, but those close to the meditator disagree. So once again, when considering BTBB interaction, the personal aspect of the interaction is a critical issue.

That personal aspect is also complicated by the wide spectrum of individual tendencies ranging from those who concentrate only on BTBB reality to those who insist on only ATB reality; from extreme Dreamers, who couldn't care less about physical circumstances to extreme Doers, who insist that only physically satisfying activity is worthwhile. Extreme Dreamers won't ever agree with extreme Doers on the proper balance for interacting with the two realities. Neither of these groups will compromise on their values. Though most of us are in the middle somewhere, with both Dreamer and Doer tendencies, we all have our own priorities and thus will constantly debate where the ideal balance is to be found.

We're not making much progress in finding that balance point, and, as already noted, we won't find one in this chapter, but perhaps the debate itself is productive. When considering the issue of balance, we experience the diversity of humanity and, in that diversity, we find freshness and growth. Meanwhile, we stumble along, spending most of our time earning a living for physical survival, but reserving some time for a form of interaction with BTBB reality. Few of us spend much time meditating, but most of us use either religion or some aspect of the arts for that purpose.

Note, however, that both religion and art can qualify for interaction with ATB reality as well—both can give genuine physical pleasure. Art can present a barrage of sights, sounds, and textures, while religion can provide a forum for festive occasions and social gatherings, such as births and weddings. The balance point, therefore, is blurred. One spouse may attend a church picnic to leverage social advantage or to pursue business interests—an ATB view—while the other may attend to achieve spiritual fulfillment by following biblical directives to cheer up the elderly or help the youth of the parish—a BTBB view.

Nevertheless, an imbalanced life is dysfunctional and problematic. Most people can recognize an imbalance, even if they don't know exactly where the problem lies. In fact, using examples of imbalance might help us to identify a proper balance. Let's try that by looking at a Doer couple.

One spouse, who's working so many hours a week that there isn't sufficient time for relaxation and frivolity, is easily identified as problematic. There may be some debate, but even though they're both Doers, even extreme Doers, the couple realizes that something beyond work is necessary for the marriage to work. So even if both spouses have the same outlook on reality, an imbalance can be identified. Once this is accomplished, the adjustment of priorities is much easier. Time for relaxation and frivolity, which is bound to include some BTBB interaction, will be found. Similarly, if a person is daydreaming and a child becomes endangered because of inattention, that person too will notice the imbalance and temper the daydreaming.

This negative approach also helps in judging leaders of society. An extreme Doer leader may try to drive a society into aggression and conflict. People must decide if that stance is legitimate or not. Of course, it's easy to agree if an enemy is about to attack. But if the threat is theoretical, based on rhetoric or posturing, then deciding the best way to respond becomes subjective and argumentative. A wrong decision can place the society in grave danger. Whether the threat is real and no action is taken, or the threat isn't real and action is taken, the results can be catastrophic. Whatever the leader decides, therefore, needs to be judged based on the degree of balance the leader demonstrates. If the leader's Doer propensity is judged to be too dominant, followers should question the decision. If that is not the case, the decision should be supported.

The Cold War provides one example. There was a legitimate threat, as citizens of Western Europe will confirm. Fighting Soviet aggression in Europe was as critical as fighting the missile deployment in Cuba. Yet European civilians fought against the deployment of missiles on their own continent which were intended to stand up to Soviet forces. Despite the loud protests, European leaders eventually agreed to the deployment, and that decision arguably helped end the Cold War. By the twenty-first century, the Cold War was over and many missiles, both Soviet and American, were removed from Europe. At the turn of the millennium, Europeans lived in a post-Cold War peace, achieved by a tough attitude that had insisted on a missile deployment many believed to be totally unnecessary at the time. Doer aggression was firmly addressed, and the result was peace.

On the other hand, fighting the Cold War in Asia with similar tactics hasn't worked. The Korean War did not solve the conflict between the North and the South, and that problem remains unresolved into the twenty-first century. Did Truman misunderstand the threat? Was he too much of a Doer? Did he ignore the Dreamer aspect of the Communist threat in Asia? Those questions aren't easy to answer and hint at the problems of coexisting with societies led by leaders holding extreme viewpoints.

If an extreme Doer or Dreamer leads a society, the society's propensity will be to wreak havoc on anyone who disagrees with it. The imbalance, therefore, is identifiable. To survive, nations must recognize the threat and position themselves accordingly. First, they must identify the type of threat they are facing, since Doer and Dreamer extremists react in different ways. Then they must counter that danger with appropriate measures. If it's a Doer threat, strength counts. If it's Dreamer, the fallacy in the idea must be highlighted and demonstrated as false to the people being misled. This technique may be impractical in some cases, but at least by identifying the type of imbalance, proper precautions and strategies can be established.

Dreamers emphasize the abstract while Doers focus on the concrete. They both recruit followers to their views, and the two points of view often divide society into factions—religious or nonreligious, academic or entrepreneurial, and the like. When carried to the extreme these divisions generate conflict and misunderstanding, but even casual disagreements can polarize a society.

Now, however, we understand the fundamental basis for each of these valid points of view, so there is no reason to suspect or condemn an advocate who disagrees with our own emphasis. Those whose opinions about reality differ from ours are simply concentrating on that aspect of reality that appeals to them. We can continue to divide ourselves into Doers and Dreamers, or, using more traditional labels, the practical and the impractical, but the division shouldn't assume that either one is better than the other. Taking such a stance would be utterly false, since the division comes from the total reality that defines our existence, the reality After The Bang and the reality Before The Big Bang. BTBB reality is so fundamental, so causal, that ignoring it or considering it as secondary or inconsequential is just as big a mistake as ignoring ATB reality, whose necessity is more obvious. Both realities are legitimate and essential for humans to fully understand their existence. To deny or condemn either would be closed-minded. No one who is an advocate of interaction with either reality is wrong. A problem only surfaces when there's a commitment to only one of the realities at the expense of the other. It's the imbalance that's a threat, not one of the realities.

Whether it's within oneself, within a family, or within a society, a proper balance in interacting with reality is important. While we haven't made much progress in determining how that balance is accomplished here, we will return to this important topic in Chapter 30. But we have established that such an imbalance is generally easy to identify. Identifying the imbalance and understanding exactly what it is will prove invaluable in developing a fuller understanding of human behavior.

As with most problems, communication is the key to resolving differences of opinion about the correct amount of emphasis on either ATB or BTBB reality. That communication is often challenging when it's between Doers and Dreamers. But if the communication is sincere and persistent, the issue of imbalance can be worked on together. Now that we understand what drives Doers and Dreamers, that understanding should help us improve such communication. Let's use what we've learned about imbalance here to analyze human behavior, first on a societal and then on a personal level.

Chapter 25
Doer/Dreamer Response on a Societal Scale

A study of human interaction with all of reality will help us recognize abnormal Doer/ Dreamer behavior as well as improve our ability to verify the genuineness of BTBB interactions. Observing differences in the perception of reality is easier on an organizational rather than a personal scale, because of the availability of multiple viewpoints that challenge and reinforce one another. This constant questioning makes any judgment less subjective than a decision made on one's own. Taking advantage of this more objective perspective, we'll analyze how societies interact with one another. We'll look for both Doer and Dreamer tendencies and see how extreme societies use their exaggerated positions when confronting their adversaries. As we examine these cultural extremes, we will also consider various methods of addressing them. Then, in the next chapter, we will perform a similar analysis at the personal level.

Societies interact with each other through their leaders, and, since followers, by definition, follow the leader's lead, they magnify the leader's tendencies. As already stated, leaders are often Doers, who are ambitious and driven toward ATB reality, but they can be

Dreamers as well. Societal interaction, therefore, includes all the risks connected to dealing with differing views of reality.

When a conflict arises, the dispute often reflects the respective leaders' outlooks, whether ATB or BTBB. Understanding the fundamental attitude of the leader, and often the society as well, is essential to a successful resolution of the problem. The affected societies need to consider **both** ATB and BTBB reality when making a decision about how to respond in a crisis. If both realities are not taken into account, then the risk of the conflict growing into war increases.

We already know that the attributes of interaction with BTBB reality are timelessness, nonmateriality, and universality. Extreme Dreamers lead with a BTBB-reality propensity and place little value on time or the physical. They use both begrudgingly, but only out of necessity. These extremists have little common ground with more practical opponents who value interaction with ATB reality, where time and the physical are crucial. BTBB extremists are confident that in a struggle with a "heathen" adversary their "higher" reality will prevail. It may take generations, but time doesn't matter. Thus, a Doer society cannot expect to win a battle using a strategy of patience and procrastination against a society led by an extreme Dreamer. The only way to resolve the conflict is for the two societies to address each other in terms that the other understands. The more extreme the societies are, the more having proper communication is essential, even though the difficulty of achieving it increases as well. Nevertheless, as the Dreamer aspect leans further and further toward the nonmaterial, and the Doer aspect leans further and further toward the material, the necessity for finding terms that the "enemy" understands grows.

We will examine some historical struggles to understand the importance of meaningful interaction between Doers and Dreamers. To begin, let's consider the Middle East conflict that has been going on for hundreds of years. This example will demonstrate how differing viewpoints and misunderstandings helped lead to aggressive action. It will clearly show how mistakes and miscalculations by both Doers and Dreamers aggravated the problems. Then we will suggest a way to improve understanding and thereby begin to construct a real solution.

Since covering the complete history of the Middle East would take volumes, our example will concentrate on recent events, beginning with the world's response to the 1979 Iranian revolution. In that year the Ayatollah Ruholla Khomeini, a 78-year-old religious leader, arrived back from exile in France and orchestrated a new Iranian government bent on extreme interaction with BTBB reality. In that government's view, ATB reality served only as a necessary aid to preserve the opportunity for BTBB interaction. BTBB reality was seen as the "higher" good; ATB reality was inferior. Islam became the foundation for all of society, and those who didn't agree with the government's interpretation of its teachings were condemned as infidels.

Doer governments looked at the new Iranian government with concern and maneuvered for advantage, trying to improve their own positions with respect to each other. However, since each Doer government had its own priorities, reaction to Iran varied, even though it came from Doers. Also, since these nations were attempting to gain material advantages, they virtually disregarded the Dreamer aspect of the situation other than to feign deference if they thought doing so would benefit them. Iran, in turn, used these reactions to further its own purposes and began engaging in aggressive behavior, even allowing students to take over the American embassy and hold its staff as hostages. The Doer nations, still concerned with their own self-interest, couldn't agree on how to react to the aggression, and Iran leveraged those differences as well. Actually some Doers saw an opportunity to use the situation for their own financial gain and courted Iran to take advantage of the wealth from its oil reserves.

One Doer government was more aggressive than the others. Iraq, led by extreme Doer Saddam Hussein, saw an opportunity. Deciding that the Dreamer-led nation was militarily weak, Hussein led an attack on Iran, sending the two countries into a ten-year conflict resulting in horrific casualties.

The other Doer nations merely watched as this Doer-Dreamer battle played out. Iran was still holding American hostages when the conflict broke out. The embassy violation deepened Iran's nontraditional image and further defined it as "different" in the view of Doer nations. There was some outcry for the fighting to stop, but the Doer nations chose not to intervene. Eventually Dreamer Iran was able to push back the invading forces, and the war ended in a stalemate. The Iraqi forces hadn't been well led and the Iranian forces had employed all their resources, including twelve-year-old soldiers, to halt the aggression. The Doer proved incompetent, while the Dreamer was determined and willing to undergo immense sacrifices rather than give up its ideals.

The result? Iraq claimed "victory," and the Iranian leadership grew more extreme. The borders held, but the basic issue persisted. The costly war had solved nothing.

Examining the next phase of the conflict provides a deeper look at the effects of extreme Doer/Dreamer disputes. Doer-led Iraq, frustrated but still viable, refocused on Doer-Kuwait and attacked there in 1990. The Doers of the world reacted to this aggression far more vigorously than they had to the war with Iran. This new war was a Doer versus Doer battle. True, there was oil involved, but oil had been involved in the Iran-Iraq conflict as well. Doer nations, facing almost identical issues as in the Iran-Iraq war, understood the new conflict much better and chose to intervene. They united and quickly sent Iraq back to its borders. As a condition for peace, the Doer nations insisted on measurable concessions, which, as it turned out, Iraq failed to meet. Nevertheless, the Doer/Doer conflict between Iraq and Kuwait was much easier for them to understand than the Doer/Dreamer conflict with Iran. They recognized the blatant Doer aggression of Iraq against a Doer ally, Kuwait. The problem was defined, and the need for a clear-cut response was obvious. Of course, there were more issues involved than the Doer/Doer relationship, but the point is that Doers understand each other far better than they understand Dreamers, and that fact applies to nations as well as individuals.

It's important to note that the challenges for a Doer society in both recognizing a problem and finding a solution increase when Dreamer extremity is involved. Doers are often indifferent to any attack on a victim with an extreme Dreamer mindset. Since

Doers don't empathize with that mindset, the victim doesn't seem as real as a Doer victim does. And the more extreme a Dreamer society is in its outlook, the less the Doers understand it. Thus, when Iran was attacked, the Doer world watched. When Kuwait was attacked, the Doer world responded. When Doer extremism surfaces as physical aggression against another Doer, the issue is more easily defined and addressed. Dreamer extremism presents more of an enigma.

Meanwhile what about Iran? Its former adversary had been thrown back and defeated. Iran had gained from the victory, since Iraq's capabilities were now considerably weaker. A Doer would expect some gratitude from Iran for these accomplishments. Remember, though, Iran's Dreamers have a different viewpoint on reality. Their society is extreme Dreamer. To them, any material advantage they may have gained was nothing compared to the dismay involved in watching a Doer power conquer a fellow Muslim state. And, as far as Iran was concerned, its own issues remained festering. So rather than cheering on the defeat of the enemy that had invaded its territory a few years earlier, Iran condemned the American Doer-leadership that was responsible for Iraq's defeat in Kuwait.

Thus far we have considered two wars causing indiscriminate death and destruction, and whose outcomes led to deeper mistrust and rankling hatred. After these conflicts, the Doer leader in Iraq still schemed, and other Doer nations still maneuvered for material advantage. Dreamer Iran still wove its web, intent on achieving its own, nonmaterial goals. However, just as Dreamers understand they must eat, they also realize that Doers respect power. In order to have any influence over the Doers, Dreamers must demonstrate power as well. Iran was dealing with Doers, and, though Iran placed little value on Doer reality, it did understand that to triumph over the Doers, it had to develop its own material might, including nuclear power.

There's more, of course, since the Dreamer issues are still out there. Iran was not the only Dreamer society at odds with Doer priorities. Another extreme Dreamer faction planned and carried out an attack on the United States using Islamic doctrine as justification. The 2001 attack on the World Trade Center finally awoke a Doer nation into taking action against a bitter Dreamer enemy. The United States responded, this time refusing to be satisfied with merely pushing its adversary back to a pre-existing border as in Kuwait. America led an invasion of both Afghanistan, where the Dreamer faction originated, and then Iraq, which still hadn't met its post-war agreements. The U.S. now called for complete victory, a demand similar to the one it had insisted on to end WWII.

Thus, the American answer to the Dreamer attack was predictable, though only partially effective. A Doer nation used Doer methods and Doer values to combat Dreamer Al Qaeda forces, which weren't convinced by either. Further, the connection between the Dreamer forces in Afghanistan and the Iraqi problem was ambiguous at best. The Doer, American-led coalition may have won the initial confrontation in Iraq, but that victory didn't resolve the problem. Instead, the downfall of Saddam Hussein allowed Dreamer-led factions, which had been forcibly suppressed under the old regime, to surface. These factions turned out to be natural allies of Iran. So, rather than a Dreamer problem going away as the result of a military victory, it actually got worse. This consequence should not have come as a surprise. The fundamental problem couldn't have improved, because it wasn't even addressed.

This long illustration reveals the problems associated with Doers dealing with Dreamers at a societal level, particularly if the Dreamers are extreme. So, if the example demonstrates an incorrect response, how should Dreamer extremism be confronted?

First, the Dreamers must be identified as what they are. Then, while their view of reality must not be condemned, they must be shown that a Doer viewpoint has value as well. Both views of reality are valid, but both are subject to becoming extreme and intolerant. Next, the Dreamer leadership must be tested for genuineness, using all the tools from Chapter 22. Once that determination is made, then the extremism must be addressed. That is, the Dreamer leader must be convinced that his or her view is extreme and unreasonable. Of course, accomplishing that isn't easy. Some might say it's impossible, but it's the same challenge that the Doer has to meet in accepting the Dreamer's legitimacy. Since the Doer overcame a natural tendency to disregard BTBB reality, the Dreamer must be convinced to respect ATB reality. Fortunately followers will often be the first to notice the problems with extreme positions, and a groundswell of opposition may influence even an authoritarian leader.

No matter what, however, an argument highlighting the extremism must be put forth, because that is the only effective method of countering a genuine, extreme Dreamer. Doer values and methods won't work; physical force will only suppress the problem not solve it. All dialogue must be pursued with that fact in mind. The Dreamer, even if extreme, must be shown respect and consideration. The Dreamer's reality is real and just as critical as the Doer's reality. Once recognizing the respect, the Dreamer might be more willing to listen to the argument. This method only works if the extreme Dreamer is **genuine**, however. Unfortunately, if there is no genuineness, it may be necessary to resort to tactics such as economic sanctions or even military force.

Another example of Doer/Dreamer confrontation can be found in the dealings between the United States and the Soviet Union. This example adds the complication of Doers taking advantage of a Dreamer idea. Let's look at Doer America facing Dreamer Soviet Union, even though the Soviets often used a Dreamer philosophy for their own Doer ends.

The Communist Manifesto proposes an idealistic solution to life's difficulties. Marx's philosophy calls for a collective sharing of wealth and ignores the time dependence of market forces that Doer capitalists consider fundamental. Disregarding these market forces and relying on central planning also has a nonmaterial element, since bureaucrats theorize about material needs and develop plans instead of relying on the law of supply and demand. Soviet communism may not be as clear an example of BTBB extremism as a nation such as Iran, but the example demonstrates that such extremism extends beyond religious fanaticism.

To begin, note that the communist philosophy as practiced in the Soviet Union presented a contradiction in its outlook on reality. The communist practitioners would argue that, for them, time and matter were critical. ATB reality was what counted. But actually it didn't. To even the most down-to-earth Soviet communists, what mattered was the PLAN, since they believed that if the PLAN was followed, all difficulties would be overcome and the promises of communism would be fulfilled. The theoretical PLAN became, therefore, the end-all. It was the product of the theory of communism. The Soviet communist philosophy was an extreme Dreamer philosophy, fed by the ideal and not by the hard ATB reality that capitalism addresses.

However, something happened to Soviet communism. The Soviet Union's Doer leaders quickly defined Marxist idealism in their own terms and used the concept to force followers into their own rigid code of conduct. These Doer-leaders aggressively expanded communist beliefs to other nations, overtaking neighboring states by force. The end result was material power—a Doer goal.

Thus, there was an interesting dynamic that developed between the leaders of the Doer (America and its allies) societies and the Dreamer (Soviet) society. Although the Soviet Union didn't trust capitalist societies any more than they trusted the Soviets, as far as communication was concerned, there was an advantage over a straight Dreamer/Doer interaction. Both factions were led by Doers, so the leadership understood each other's goals with regard to political power. The result was a vast build-up of military might by both sides. In addition, the underlying issue, the conflict between a communist Dreamer ideal and a capitalist Doer ideal, remained.

So when dealing with the Soviet Union, there were two issues: the Doer leadership and the Dreamer philosophy. Addressing the philosophy required the same strategy as addressing post-revolutionary Iran. Addressing the Doer leadership required the strategy that works for Doers—force, either actual or implied. That is, dealing with the Soviet Union meant not only dealing with an idealistic doctrine, but also required sufficient military capability to meet an aggressive Doer-stance.

Looking at Soviet communism shows that defining true reality for competing societies can be difficult, since opponents may not represent their position frankly. Doers may be using Dreamer political values as a tool for their own practical ends. Doer leaders may also take advantage of religious beliefs to motivate followers. Dreamer leaders may think they're dealing with another Dreamer, when actually the opponent is a Doer merely using religion. The twelfth-century Christian Crusades are one example of this eventuality. Despite the problems involved, identifying the actual motivation of the opponent is critical.

Note that for the U.S.A. and the U.S.S.R., while the Doer leaders understood each others' Doer motivations, the functionaries of each society, the capitalists and the communists, found communication far more difficult. That is, although American politicians understood the Soviet use of power, American capitalists didn't respect the Dreamer reality that lay at its foundation. Communication between the two societies suffered accordingly. To deal with capitalist Doers, communist Dreamers needed to take into account the Doer philosophy that emphasizes material wealth and time. But it was repulsive for Dreamer communists, who saw the market as capitalistic greed tromping on the masses, to reduce themselves to the profane level of marketing. On the other hand, doctrinaire capitalists, for whom market forces are primary, saw the PLAN as nonsense. Thus, relationships between the two societies encountered one difficulty after another.

What worked? First, communist aggression was easily identified as Doer-aggression. In the end, the massive military might in the West proved to be superior to the military might of the Soviet Union. However, what solved the underlying issue was the internal collapse of the Dreamer-society. It simply couldn't support the ATB reality that the aggressive Doer-leadership was pursuing and was using communist theory to achieve.

Long before the actual collapse, many American Doers had condemned the communist economic system as being unable to function in the real world as effectively as capitalism. But if those same proponents had made the distinction between BTBB and ATB reality while demonstrating the shortcomings of communism in dealing with the ATB world, the problem might have been resolved sooner. That conclusion may be wishful thinking, but once people realize the distinction between the two realities, there is more opportunity for making progress. Conflicts and tension won't just disappear, of course, no matter how great the understanding between competing parties. But adding a clear grasp of total reality to the equation can't help but be beneficial.

The above examples are typical of the history of Dreamer/Doer conflict. When misunderstandings grow, armed violence is often the result. Instead of war resolving the issue, however, the problems continue to ferment, leading to a cycle of hatred and revenge. Combating hatred with massive, overwhelming force may suppress the conflict, but doesn't eliminate it. Generation after generation may wait for the opportunity to obtain what they define as justice. Much of the world is enmeshed in this type of battle in one way or another. Lands are redistributed and a new "peace" is proclaimed, but the peace cannot last unless the fundamental issue is addressed. And that issue includes resolving any differences in interpretation of what true reality is. Doers look at Dreamers as impractical idealists, while Dreamers look at Doers as rabid materialists. But true reality is a balance of **both** of these two positions.

The conflict between those who advocate the sacred and those who choose the profane, i.e. spiritual vs. secular, is so basic that it is sometimes addressed in the fundamental institutions of society. The American Constitution is a good example. America's founders understood the danger of combining the power of the state with the teaching of religion. At the time, European religious leaders were often either Doers using religion as leverage for power, or Dreamers who defined reality narrowly in the BTBB direction. These American fathers established a document that separated the functions of church and state. The new society insisted that differing points of view be tolerated and that practitioners be allowed to worship as they wished. There would be no governmental role in religion. Thus, the American experiment was an attempt to avoid the conflict of Dreamer and Doer extremes by separating the more formal elements of each, while emphasizing the legitimacy of both. Proper communication and tolerance allows Doers and Dreamers to complement each other rather than view their opposites as propagators of evil. Whether inclined toward the practical or the nonmaterial, under the American system, one can communicate with neighbors of different faiths and live in harmony with them.

Although this strategy allowed opposing viewpoints to coexist, it didn't entirely resolve the issue. No matter how well laws are written, the spiritual (BTBB reality) keeps creeping into everyday life, just as the physical (ATB reality) creeps into religious practice. Should auto dealers sell cars on Sundays? Should liquor be sold? Should it be sold on Sundays? What about faiths that consider Friday or Saturday the holiest day of the week? Despite continuing debate on these questions, however, the separation of church and state promotes harmony. It may be an incomplete solution, but it facilitates understanding by regarding all views as equal.

In summary, Doer/Dreamer issues involving opposing views of reality have been in existence since the very beginnings of civilization. For much of that time, mankind has thrown up its hands and relied on force, or the threat of it, to resolve any differences. While there has been progress, much of that progress has been made by skirting the issue rather than addressing the cause. Armed with new knowledge, we can now approach the issue more fundamentally.

To combat an adversary with a different view of reality, the threat must first be recognized for what it is. That is, the type of leadership must be established. If the leader is a Doer, like Saddam Hussein in 1990, the identification is easier, and the required response clearer than for a Dreamer. Yet the implications of any response must be well understood as well, since Doers like Hussein may have repressed extreme Dreamers who will surface at the first opportunity. Moreover, there are many other scenarios where the identification itself isn't simple. For example, the leader may be a fanatic Doer using Dreamer philosophy to gain power and influence, as in the Soviet Union, or a Dreamer fanatic out to destroy Doer reality, as in Iran. In either of these cases, the leadership uses interaction with BTBB reality for its own purposes. If Dreamer leadership is extreme enough, it will advise its followers to avoid any unnecessary interaction with ATB reality. Thus, it will condemn the Doer mentality. It may even warp BTBB reality to achieve its goals. Such perversion could include physical violence, where any consequences to the innocent (such as "martyred" followers) are considered noble and acceptable sacrifices. In that case the Dreamer would be using Doer tactics, since life itself is ATB reality. All these variables make it difficult to identify the actual reality that a threat represents. Also, the analysis must include an estimate of genuineness and a clear view of both the leader's and followers' outlooks on reality.

If the leadership, whether Dreamer or Doer, is also identified as fanatical, it's usually possible for an objective evaluation to demonstrate how blatantly false its reality is. The problem, of course, is for the leader and followers to accept that evaluation. Since Dreamer fanaticism doesn't respond to physical persuasion, the only way to fight it is to learn as much as possible about the philosophy and react accordingly. The real battle is fought using images, not violence. When dealing with Dreamer fanaticism, any victory won by force is temporary, since, unless the movement's philosophy is discredited, converts will continue to give up their lives no matter how impossible the odds. Combating a Dreamer philosophy using procrastination or force may give rise to a peace that lasts for a few years or even for a few generations, but eventually the believers or their descendents will reemerge. The philosophy itself must be refuted and shown to be invalid. Once the fallacy is clearly identified, its destructive nature must be trumpeted as widely as possible, with the hope that the message will penetrate the society.

Another mistake in combating a fanatic Dreamer enemy is using Doer tools as punishment. For instance, a sentence to a long term in prison has little meaning for a Dreamer. A better sentence would be structured interrogation with the purpose of learning as much as possible about the movement, its leaders, and its followers. Time would not be a factor in the penalty. The detainee would be held until all valuable information had been gleaned, and the detainee had demonstrated absolute conviction in the fallacy of the cause. Incarceration could last from a day to a lifetime. The Doer American society would probably consider such a proposal as outrageous. Nevertheless, when dealing with

Dreamer fanatics, Doer values and rules don't work. Instead, a proper communication conduit must be found.

Another way to fight fanaticism effectively is to interrupt the recruiting process and redirect the recruit's thoughts toward a balanced view of reality. Surprisingly the redirection, although involved, may not be as difficult as it seems. The warped view of reality fanatical philosophies represent is usually obvious. First, it's extremely unbalanced, so unbalanced that human life is considered disposable. Second, the interaction with BTBB reality is illegitimate. Fanatical Dreamers insist on BTBB interaction that's void of ATB reality—that's the definition of Dreamer fanaticism. Third, its reality is far from universal. These philosophies claim special, specific advantages that make their followers superior to the enemy that they've defined. Of course, recruits will be unaware of the implications of imbalance, legitimacy, and universality in the way we have addressed them. However, once these ideas have been explained properly, and once the recruit has regained a balanced view of reality, their importance will become apparent. Opening the recruit's mind to accept such an explanation may prove challenging, but redirecting that mind is essential.

While these methods are the best way to deal with fanatics, criminal or non-genuine fanaticism can't be overcome by communication. It didn't work with Hitler, and it won't work with any other criminal. Although war is repulsive and must be the last resort, to ignore or to tolerate the threat can lead to even more catastrophic results. However, knowing the distinction between the two aspects of reality, ATB and BTBB, and addressing that distinction may help avoid such calamity.

Learning how to determine whether interactions are Doer or Dreamer in a fanatical society makes it easier to identify extremism on a more personal level. So let's turn to the individual, someone you may know and love. What happens if a spouse or offspring or parent doesn't have the same outlook on reality as you do? How does that affect your communication? What happens if your view or your spouse's view of reality changes? How should you react to that change? Worse, what happens when the outlook becomes extreme? We'll address all of these questions in the next four chapters.

Chapter 26
Doer/Dreamer Response on a Personal Scale

When analyzing individual differences in interacting with reality, we'll discover a range of outlooks, from Pseudo Dreamers at one end to Manic Doers at the other. Pseudo Dreamers play with BTBB reality rather than interact with it legitimately. Manic Doers, on the other hand, are so dedicated to ATB reality that they use whatever means is available to advance themselves in that reality—at anyone else's expense. Of course, the vast majority of people lie somewhere in the middle. However, we'll see that that middle includes extreme Dreamers, who dedicate themselves to the abstract, and extreme Doers, who often become successful leaders in a capitalist society. We'll differentiate these individuals from the Pseudo and Manic versions and thereby gain a clear view of the full range of outlooks. Whatever our own propensity, we'll learn that there are many others and see that these differences can become a source of human conflict.

We'll begin our analysis of individual preferences by looking at two familiar relationships where differences surface, generational and marital. Then we'll investigate the abnormalities that emerge when frustrated individuals are prevented from interacting with the reality of their choice. Let's start with generational issues.

Children can drive their parents crazy, especially when they decide they are old enough to make judgments about behavior and manners on their own. As they mature, they tend to question and even ignore parental values in favor of new ones instilled by their peers, the media, and/or the opposite sex. They conclude that their parents are old-fashioned and don't understand the modern world, even though they may have no idea what "modern" actually implies. The battle commences and doesn't end until the child either learns from experience or actually does mature and begins to look at innovation with an appropriate dose of skepticism.

While this generational battle is common and usually temporary, it can escalate into a serious and permanent condition. Although the child may be merely displaying a new behavior with traits that reflect a new "style," that change could also indicate a differing view of reality. The child may be developing into a Dreamer. A son's sloppiness of manners can signify more than mere rebellion against authority. It may mean that he has decided that any time spent on something as mundane as cleaning is wasted and could be better spent on music or poetry. He may have latched onto an idea, philosophy, or political view that is contrary to his parents' way of thinking. Perhaps he has experienced a religious conversion and intends to apply for membership in a community whose spiritual values intrigue him. Or a daughter may choose the opposite direction and become a Doer. She may quit church, quit band, quit anything that emphasizes the abstract or artsy aspect of life. Instead she will insist on pursuing the **real** reality, one that leads to **real** achievement such as financial success or athletic championship. In both of these cases, frustrated parents throw up their hands and wonder what they did wrong. Why would a child suddenly go against principles that the parents had so carefully instilled? Religious parents can't imagine a child becoming an atheist, just as secular parents can't imagine a child leaving home for a monastery.

Spousal differences in viewing reality are another source of frustration that can lead to conflict. In love, opposites attract, and marriage often takes place between two people who complement each other. What could be a more fitting pair than the combination of a Doer and a Dreamer? Each observes a refreshing, alternative outlook on reality in the other, and the decision to marry appears obvious and natural. Initially the relationship may thrive as anticipated. The Dreamer enjoys the Doer's feel for the practical, and the Doer finds the Dreamer's insight into the abstract fascinating and fulfilling. As time passes and the novelty wears off, however, tension can surface. The Doer may voice complaints about the Dreamer's lack of effort and ambition. While the Dreamer may respond sympathetically, a true Dreamer will continually remind the Doer that there are roses to smell, songs to hear, and art to enjoy. Even though the Doer may agree, the gnawing on the Doer's psyche will continue, and, eventually, if the Dreamer fails to respect the Doer's view of reality sufficiently, the Doer may conclude that the situation is hopeless. Of course, the same applies if the Doer refuses to participate in the Dreamer's reality.

To achieve a successful relationship, both the ATB and BTBB reality must be given ample consideration. Even extreme Doers have some Dreamer in their psyche, just as extreme Dreamers have some Doer. Still, since one of the attitudes predominates in most people, agreeing to the premise that both realities are equally important isn't easy. Think of the Doer trying to cope with a spouse who demands expensive piano lessons when the roof is leaking, the car needs new tires, and the children's college fund is zero. Or—in the case of rebelling offspring—think of the Doer parent coping with a child whose listening to loud, amplified music supersedes all other activities, including homework, chores, and family conversation. Challenging the behavior doesn't help. The child sulks and withdraws deeper into the music, exacerbating the problem. The Doer parent panics. "Turn off that noise! Go out there and mow the goddamn lawn!"

For a family to succeed, there must be an understanding that two realities exist and that they are of equal importance. The parents, whether Dreamer or Doer, must insist on sufficient attention to both realities for all members of the family. The Dreamers must learn to concentrate on everyday reality, and the Doers must leave time for the abstract. Both the Doers and the Dreamers must go more than halfway in meeting the other's preference for interaction with reality, and, if all comply, the family will take on a new and more exciting texture. Of course, that means that Doers must force themselves to appreciate music even when the house needs painting, the laundry needs washing, and a kitchen chair needs repairing. And Dreamers must force themselves to concentrate on the practical, even if it means missing an important art show. This accepting attitude applies to all relationships, husband with wife, parent with child, and friend with friend.

The above advice goes far beyond insisting that the two sides compromise and merely agree to accept each other's outlook on life. The issue is much deeper than taste or opinion. The issue is how to deal with reality, all reality, both ATB and BTBB. If either is ignored, there are serious consequences. Timelessness and nomateriality count, just as food and warmth do. A given preference for reality may work on a personal level, but when others are involved, that preference must be tempered, and the tempering will give all concerned a more balanced perspective.

However, just as both realities need to be addressed by all, care must also be exercised when insisting on the amount of another's interaction with either. Such insistence can lead to the repression of natural inclinations and seed a strong, negative reaction that will germinate and grow later on. That is, when individuals are prevented from interacting with their reality of choice, they may take drastic measures to overcome the obstacle. These measures can include turning to alcohol or drugs, and, once addiction has polluted the situation, the path to compatibility and success becomes far more complicated.

Consider a Dreamer addict. The addiction may begin because too many barriers have been placed in the Dreamer's way. As a result, the Dreamer attempts to create BTBB interaction through illegitimate means. However, drugs bring physical dependency, the very thing the Dreamer is trying to flee. Rather than achieving independence from the physical, the addict is forced to respond to physical demands that damage the body and consume resources (which tend to be meager for a Dreamer anyway). Thus, there exists a dichotomy between a physical dependence and the need for interaction with the reality that the Dreamer finds more appealing. Once hooked, the Dreamer is caught in an ever-deepening vortex.

But there is help. Illegitimate interaction with BTBB reality using mind-bending chemicals is easy to identify. Society has had sufficient experience with this malady to establish a variety of methods to combat it. Tools are available to assist in overcoming addictions: medical therapy, such as methadone; religious conversion, such as born-again Christianity; or social/religious networking, such as Alcoholics Anonymous. While these tools can be effective in combating the problem, however, they may not be a cure.

Now we have a new tool. Its use may not eliminate physical addiction either, but it will help convince an addict that the drug must be avoided as one avoids anything poisonous. This new tool involves persuading the addict of the importance of total reality. If the idea is properly presented, the addict will recognize the drug **as** poison, and, thus, in a sense, even the physical addiction may be cured. This method differs from those that concentrate on the addiction itself. For instance, some Doers attempt to treat Dreamer addicts by claiming that the drug use was caused by the desire to escape from reality, and, as a consequence, condemn that desire as well as the drugs. Thus, the equivalent of any interaction with BTBB reality is forbidden. Such treatment could be disastrous, especially if the craving for BTBB interaction is the fundamental cause of the problem. Nurturing that craving is far more important than condemning the drug. It's more effective to point out that the drug doesn't legitimately satisfy the healthy and normal desire the addict has. Drugs only provide a temporary and false satisfaction.

To treat an addict successfully, he or she must learn to properly respect all of reality. That is, the addict must develop a firm conviction that interaction with ATB reality is just as critical and just as important as interaction with BTBB reality, and only genuine interaction is lasting. To accomplish this, the addict should receive a healthy dose of interaction with the practical and an equally healthy dose of interaction with BTBB reality. The stronger the dose of both, the better. Also, there needs to be a method for balancing the two interactions properly. We'll get to that method in Chapter 30. Let's note here, though, that the addict must learn to value a complete life that includes both realities, a life where both ATB and BTBB reality are natural and unforced.

Of course, addiction isn't the only situation where people have problems understanding someone else's view of reality. Just having an extreme position, whether it be Doer or Dreamer, causes difficulty. But the solution is similar. Once they learn to interact with the other's reality, extreme Doers and Dreamers can relate more effectively to each other. However, merely suggesting such interaction isn't enough. For the extreme Dreamer, a propensity toward a reality where the physical doesn't count and time isn't important usually overrides any advice from a Doer, especially since the Doer's cold logic is grounded in ATB reality. That's why using activities to promote ATB interaction is preferable. Even something as simple as playing games can help, the more physical the better. Instead of bridge, use touch football—as long as it's not played too competitively. The Dreamer needs to feel the joy of ATB reality, not the pain that often accompanies it. Once the pleasure involved in engaging in such activity is established, an argument can be made advocating regular interaction. With constant reinforcement, the Dreamer may eventually be convinced. This technique could help an addict as well, but the physical dependence of the addiction interferes with and complicates the process.

Extreme Doers can be dealt with in a similar fashion. An argument expounding the advantages of BTBB interaction will most likely be futile. However, an easy, pleasurable experience with BTBB reality such as attending a good play, preferably a comedy, can help. Again, with sufficient persistence the extreme Doer might eventually agree to a shift in priorities.

The secret for success in helping extreme Doers and Dreamers understand each other is to get them to experience the advantages of interaction with both realities. The Dreamer certainly understands the need for nourishment, and the Doer will laugh heartily at a good joke. Although Dreamers hate repetitive labor, they often enjoy its fruits. Similarly, Doers may avoid the art museum more out of impatience than a lack of appreciation. Deep down, reasonable Dreamers and Doers appreciate the other's choice. They may even envy their opposite.

Another point to emphasize is that extreme Doers and Dreamers must respect each other's point of view. That is, they must present their preference for reality in a manner that's not insulting or degrading, as well as being willing to listen attentively to their opposites' perspective. This isn't easy. For an example of this situation, we'll use the conflict between an extreme Doer father and his Dreamer son who's experimenting with marijuana.

After spending a long day in a profession that requires total commitment and all the energy a father can muster, it's almost impossible to come home and be sympathetic to a son found smoking pot. As with the addict, the drug use may be a symptom of frustration in a child who sees no opportunity for interaction with BTBB reality. He mistakenly believes that the interaction can be found in a drug that removes everyday reality or at least sends it to the background. Typically, the father is blind to the basic problem and instead concentrates on the symptom. The father demands to know if the son is a habitual user. Immediately the son senses both hostility and insincerity in the tone of his father's voice. As far as he is concerned, his father has already condemned him. Besides, he's only smoking. He prefers marijuana to tobacco. So what?

The son responds that marijuana isn't additive like tobacco and stomps away. The father explodes in a rage, remembering the nicotine patches and other gadgets he depended on to kick the cigarette habit. Thus, what began as a well-intentioned hello, deteriorates into a verbal hodgepodge of insults, accusations, and condemnations from both sides. The father knows that the son doesn't have an adult's understanding of life but can't figure out how to explain the problem in a way the son will accept. Meanwhile, the son's perception is confirmed: the father is insincere, closed-minded, and prejudiced. So what should the father have said? Remember, the kid's on drugs, for God's sake!

The father must convince the son that any judgment he makes is sincere and respectful. To get to that point, the father must first learn what led to the use of marijuana. If the son is trying out a drug because there's a void in his life, it's the father's job to find out what that void is. Why is the son so frustrated that he turns to something artificial to ease his pain? By showing sympathy for the son's attitude, the son can be made to see him as a helper rather than a judge.

The important thing to remember is that however the issue is addressed, the advice must encourage the son's fundamental desire. It must emphasize the positive aspects of that desire rather than the destructive aspects of the drug. The healing can begin once the son understands that his preference is not only normal but perfectly healthy. If the son is leaning toward a Dreamer extremity, he will eventually need to learn that artificial means will destroy the body instead of satisfying the craving for BTBB interaction. However, save that lesson for last. Only after the child realizes that his craving is healthy and natural, should the destructive aspect of artificially induced satisfaction be brought home. Eventually, the father and son may agree that the father has the opposite problem. If there has been too much emphasis on interacting with ATB reality, the father may decide to change as well. Thus a positive conversion on both sides can commence. Although there may be setbacks, if nurtured properly, the son will reach a healthy balance of interaction with both ABT and BTBB reality.

Deficiencies in interaction with one of the realities as well as illegitimate interactions are a major source of human strife. It is essential to recognize the need for interaction with both ATB **and** BTBB reality on a genuine rather than an illegitimate level and find an appropriate balance between the two. In this chapter we've shown how a personal preference for dealing with reality can cause conflict between individuals. The problem is exacerbated if the personal preference is frustrated, and/or if artificial means are used in an attempt to induce BTBB interaction. In the next chapter we will begin to identify extreme individuals, look for the reasons for their condition, and observe the ways in which they affect society.

Chapter 27
Dreamer Extremes

Excess can be a problem in any human activity, and interaction with ATB and BTBB reality is no different. Such interaction can be so heavily weighted in one direction that it becomes disruptive. In this chapter we will look at Dreamers and the ways in which their overzealousness can affect society, and in the next two chapters we'll move to the Doer end of the spectrum. For Dreamers we'll begin with a brief overlook, observing their role in society and how extreme behavior distorts that role. We'll consider the influence of the Dreamer viewpoint on an individual's attitude, particularly during childhood, and then give an example of a Dreamer deteriorating into the most extreme form, the Pseudo Dreamer. Following a description of this condition, we'll suggest a method for its cure. So let's give the Dreamers, those who focus on BTBB reality, our attention. If you're one of them, enjoy your turn in the spotlight, because it will be one of the few times most of you will get such notice.

Dreamers can exist in any society at any position. There are Dreamers who work in fields, in factories, and in hospitals. There are Dreamers who stay at home and Dreamers who spend most of their time in bars or clubs. There are Dreamers who watch parades, and there are Dreamers who march in parades as well. What is it that distinguishes Dreamers from the rest of us?

Since Dreamers prefer to interact with BTBB reality they love the abstract, the theoretical, the fanciful, and the miraculous. If they work in a farm field or factory, they do so mechanically, their hands performing the assigned tasks but their minds engaged elsewhere. If they work in a hospital, they concentrate on patients' hopes rather than their diseases. They will treat the disease, of course, using their expertise as effectively as they can, but treatment itself won't be their forte. Dreamers are the medical workers with great bedside manners. They lift spirits and instill confidence.

If Dreamers stay at home, they ignore cobwebs and messiness and concentrate on what might be. If they sit in a bar or at a club, they think about other times and other places. If they write, they tend to prefer fiction over nonfiction and poems over prose. The more freedom from rules, the better. Dreamers couldn't care less about the here and now and couldn't care more about the could-be and the what-if. As a consequence, most Dreamer households are financially challenged.

When a Dreamer concentrates too much on BTBB reality, ATB reality suffers. How much depends on how extreme the behavior is. If a Dreamer daydreams while driving and doesn't pay attention to traffic, the result could be a tragic accident. Most of the time the aberration is merely inconvenient, however—a missed appointment, a forgotten purchase, or the late payment of a bill. While the result may be bothersome, no lasting damage is done.

Significant issues can arise, however, when parents and children do not share the same outlook on reality. Whether it's Doer children with Dreamer parents, or Dreamer offspring with Doer parents, the situation is often problematic. And in either case, the plight increases as the extremity of the Doer/Dreamer outlook intensifies. So to begin our analysis of the impact of Dreamer extremity, we'll look at a Doer who is an offspring of Dreamer parents and then at a Dreamer offspring of Doer parents. In the latter example, we will show how the Dreamer offspring can become so obsessed that the most extreme case, the Pseudo Dreamer, emerges.

Consider Harv, a Doer who had Dreamer parents. Harv loved contact sports, the harder the hitting, the better. For some reason, each impact fine-tuned his senses and made him feel more alive. When he was a freshman in high school, he was eager to make the football team. His Dreamer parents, both teachers, discouraged the idea. They talked about knee injuries and pulled muscles. What's wrong with concentrating on music? they asked. Try out for the school jazz band instead. Thus, his parents pointed him toward the theoretical and the ideal, which, for Harv, meant the intangible and the unlikely

Harv wasn't dense, and he detected a note of hypocrisy in his parents' attitude. He was aware of the risks involved in playing football but also realized that when he volunteered to help with a religious group in a dangerous neighborhood his parents applauded his bravery. As far as making the team was concerned, injury wasn't the issue; sports was. In any case, that group volunteer program never accomplished anything except to spruce up an old woman's porch. And when they were through, she told him that she never even used it. Too many bugs! Harv wanted to see immediate, positive results for his efforts. Perhaps the woman's neighborhood was improved after the clean-up, but when his group went back a month later, the area seemed just as messy as before. No, the volunteer work was a waste of time, no matter what his parents kept telling him.

Kids like Harv don't appreciate the foggy targets that Dreamer parents give them. There is nothing to grasp, nothing to strive for except a lofty goal of abstract or artsy accomplishment. A neighborhood clean-up for strangers who are indifferent to the process doesn't present a clear-cut goal for them. It may be fine in theory, but Harv isn't into "theory."

In addition to fuzzy goals, Dreamer parents may offer an obscure path for achieving those goals. That is, while they give rules, merely following those rules may or may not be sufficient. In some cases violating the "rules" is required. It certainly was for the jazz band that Harv's parents insisted he join during the summer of his volunteer work. The leader kept telling Harv to improvise, but Harv had no idea what notes he should play, if they weren't the ones written down. All the complexity of the abstract made it tough for Harv to find his way.

But that's Harv's viewpoint. Think of his Dreamer parents. They were raising a child who was challenging their outlook on life. Although they realized that teenagers often question their parents' priorities, they couldn't ignore Harv's attitude. His football quest was just the latest issue. He'd always been too materialistic. He'd hated music ever since his first lessons when he was four. What was his problem? Why couldn't he see the value of art and thought? Why didn't he study harder? He was smart enough. Why didn't he make the honor roll?

Thus, the conflict between the Dreamer parents and their Doer son will continue, and it won't be resolved until they understand each other's viewpoint. Of course, the extremity of that viewpoint is also a factor. The more insistent the parents are on having Harv interact with BTBB reality, the more difficult it will be to reach any agreement.

When the child is a Dreamer and the parents are Doers, the issues may be different, but no less difficult. Then the Doer parents will complain that the child isn't practical. For a daughter, that practicality often shows itself in social popularity and extracurricular accomplishment be it in sports or clubs. If the daughter is unpopular or bashful, the Doer parent will offer encouragement and suggestions, anything to get the girl to focus on immediate ATB reality. Any parental advice on academic affairs will likely focus on a professional career, not on improving the daughter's appreciation of the fine arts.

For a son, Doer practicality frequently shows itself in athletic competition. If a son doesn't make one of the athletic teams in his high school freshman year, Doer parents will often show concern, wondering what's the matter with the kid. Why didn't he even try out? Most Doer parents are persistent and will search for a place where the son's talents can shine. Even if it's in something impractical, like music or art, there is always the hope that he'll win a scholarship. Thus, unlike Harv's parents, Doer parents provide their offspring with very defined goals. The problem is that those goals don't necessarily fit in with the child's dreams.

Doer parents can be so insistent on imposing their views that their offspring repress their true outlook until they have enough freedom to embrace it. This often occurs in college, where the child has the first opportunity to strike out independently. It may not happen right away, but eventually the hidden Dreamer mentality comes to the surface. It can even grow into an extreme Dreamer outlook. A seemingly bright and talented student, studying a discipline designed to train the mind for a rigorous task, will suddenly abandon that pursuit and go off on an unexpected tangent. The student begins to spend all free time in the library, wandering through aisles filled with poetry, literature, and music.

The change is easy to spot, of course. The "converted" student has made a new choice for reality, and a bookworm mentality kicks in. Doer peers fret that they've witnessed a destructive change in their classmate, convinced that the new choice will lead to a wasted life. The convert ignores those views as well as any negative feedback from Doer parents. For this Dreamer there's finally a chance to interact with the reality of choice. All that matters is the other reality that now has the student's full attention.

While the student may eventually regain some balance, that isn't always the case. He or she may become more and more engrossed in BTBB reality—so obsessed that the student becomes a Pseudo Dreamer.

Pseudo Dreamers demonstrate illegitimate interaction with BTBB reality in that they don't value the embedded ATB reality that's necessarily a part of all human interactions. They attempt to immerse themselves totally in BTBB reality. They may evolve into their pseudo state as the result of an unsettled childhood, or they may reach that state on their own, due to social inadequacies, economic forces, or other external pressures that they can't handle. Once in the pseudo state, however, they make a mockery of a sincere quest for interaction with BTBB reality. There's no genuineness in their interactions, since they violate all the rules for genuineness. First, they totally ignore ATB reality. Second, they don't try to explain their ideas using normal, everyday concepts, because they're sure that they've advanced beyond such plebian thought. Third, there is no real nonmateriality, timelessness, or universality in their interactions.

Pseudo Dreamers are everywhere. There are Pseudo Dreamers in politics, where verbose, meaningless chatter grabs votes; in entertainment, where fantasy and reality merge to blur real achievement; in big business, where office gamesmanship dictates promotion; and even in science, where the thirst for discovery can override the legitimacy of an experiment.

Let's go back to the student and see how he or she turns into a Pseudo Dreamer. Remember, the student is intoxicated by the new freedom to indulge in a different reality. BTBB reality becomes more and more predominant, and the old social network disappears. Peers leave school, marry, and move on. Still intoxicated, the Dreamer has no desire to leave. Rather, the student is anxious to explore new intellectual fields and collect new graduate degrees, none of which lead to a career. All of the coursework is undertaken merely to pursue the new reality.

Schools, rather than encouraging termination, frequently offer a small stipend that keeps the student fed and housed, since the student now supplies the school with an inexpensive instructor. So the Dreamer lingers. Perhaps beyond the age of thirty. Beyond forty. Perhaps never marrying. The rest of the student's life may be devoted to absorbing facts and appreciating the work of others. This isn't necessarily bad. That is, there's no harm to the school, fellow students, or the community. On the other hand, it isn't necessarily good either, since the potential of the student may be buried under the debris of abstraction.

Unfortunately, the imbalance can keep growing. In this case, for example, the proclivity for interaction with BTBB reality might grow so strong that there wasn't any actual BTBB interaction at all. If the Dreamer became so enamored with BTBB reality—with abstract thought—the thoughts themselves might not be real. That is, they would be non-genuine ideas blurted out simply to impress listeners. We know that interacting with BTBB reality must include all the criteria we've set forth. In this case, the timelessness, nonmateriality, and universality of genuine BTBB interaction wouldn't be there. Instead there would be "theories" composed of untested and unexplained ideas that weren't legitimate. In this warped view of reality, the individual would believe that the equivalent of pure interaction with BTBB reality would not only be possible, but frequent. The student might even disintegrate into mysticism and cult. Thus, this radical Dreamer would now have only a false interaction with BTBB reality and would become a member of the most fanatical set of Dreamers— the Pseudo Dreamers.

For Pseudo Dreamers total reality is skewed and unworkable. Fortunately, since their aberrations tend to be harmless, society treats these individuals with begrudging tolerance. Who can complain about someone who simply has an abnormal proclivity toward learning or contemplating the abstract? Society values education, and very abstract education allows mankind to reach above the practical and fulfill a deep-seated desire to address the finer points of existence. This desire is actually the appetite for interaction with BTBB reality that we all possess. When an individual embraces that appetite to an extreme, society may be curious, but it won't necessarily object.

Of course, society won't reward the individual with financial gain as it would reward a great inventor, but it may support the individual's existence with artistic opportunities, educational grants, tax-supported playhouses, and the like. In order to sustain life, Pseudo Dreamers usually require some such form of support. If they don't get it from family, academia, or the government, they often look to the wealthy, the generous, and the powerful. Thus, Pseudo Dreamers are a concern to society because they become expert at sucking up its resources.

Pseudo Dreamers resist any sincere attempt to help them, since they feel naked once the phoniness of their reality is exposed. Any attempt to challenge them into interacting with ATB reality is met with resistance that borders on aggression. Attacking their motivation is equivalent to convincing an athlete that muscles don't count. Muscles **do** count. My theory **does** matter. How dare you challenge my sincerity! These unfortunate individuals have no idea that they're using a true reality, BTBB reality (which they probably wouldn't understand and would have no desire to understand) to mask their personal inadequacies. If society attempts to discredit them, Pseudo Dreamers move on and hastily leave any community that reveals them for what they are.

The only way to force Pseudo Dreamers to come to their senses is to challenge them repeatedly with ATB reality. Support must be cut off immediately when the malady is detected, despite the discomfort it may cause. For example, if a loved one has this problem, the fastest cure may be to sabotage an application for a grant or to report incompetence to a supervisor. Such an act will draw a hateful and disowning response from the Pseudo Dreamer, of course. The Dreamer will understand the threat and feel the intense pain of insecurity, showing that withdrawal from a nonphysical addiction can be almost as torturous as withdrawal from a drug addiction. The pain is well worth it, however, if it forces the Pseudo Dreamer back to a balanced reality.

Another way to help someone tending toward this condition is to find an older person who's already a Pseudo Dreamer and show the result. While the elderly Dreamer's situation won't exactly match your acquaintance's circumstance, the similarity may be enough to convince your target to alter his or her outlook on reality. It's worth a try.

Of course, just as Dreamers can exaggerate their preference to an unhealthy extent, so too can Doers. In the next chapter, we'll go to the other end of the spectrum and look at fanatic Doers. We'll see what characteristics individuals at that extreme possess and how we can identify them. Eventually, we'll work our way to the most extreme Doer example, the Manic Doer. Just as with the Dreamers, by observing Doer extremes we can learn from their aberrations and obtain help in identifying both genuineness and imbalance. We will thus gain new expertise in determining legitimate and balanced interaction with both realities.

Chapter 28
Fanatic Doers

Doers prefer to interact with ATB reality, and, as we've pointed out, their preference is perfectly normal and acceptable. In fact, for the most part, Doers make society work. They provide the necessary materials and services that allow others to function. Society appreciates their efforts and rewards them accordingly. The incentives to be a Doer are great, and few people ignore those incentives. A problem occurs, however, when Doers take their proclivity to the extreme. Fanatic Doers interact with ATB reality to such an extent that they ignore BTBB reality almost entirely. They are totally absorbed with the practical and can be relentless in their material pursuits.

American society is a Doer society and, as such, offers plenty of opportunity for its members to become extreme. We will spend considerable time in this chapter noting how this can happen when raising children. We will then analyze the results for society as a whole as well as for individual circumstances. Finally, we'll study a fictional example of a marriage destroyed by extreme Doer mentality. In the following chapter we'll take this problem to the next level—Manic Doers.

Doer fanaticism proves to be more challenging when raising children than Dreamer fanaticism, partially because it's more difficult to recognize that a problem exists. An early hint might be poor performance at school, the result of a student deciding that studying is too abstract. Many Doers flunk out of school, especially higher education, because of a belief that what's being taught has no practical use. Occasionally these same individuals become astonishingly "successful" as a Doer society defines success. They may have been dropouts, but these Doers can turn into ruthless business entrepreneurs, scrambling to the top, and destroying anything or anyone in their way. Seeing this aberration early on, however, is difficult. Most parents will excuse a Doer son who abandons Chaucer for football practice. (On the other hand, think of the parent of a Dreamer child who abandons football practice for Chaucer. That child is far more likely to be suspected of having a problem with balancing reality.)

A young Doer is often the offspring of Doers, so the child's inclinations are encouraged. A wink and a nod, and the child understands. The child grows up, never treated for the symptoms of imbalanced interaction with reality. The practical, ATB reality rules supreme. Society nourishes the tendency, showering the Doer with material rewards and the public honors that go with them. Thus, Doer fanaticism can be nourished throughout childhood.

It begins at birth. Holding a newborn may be the warmest, most heartfelt feeling of human existence. Myriads of thoughts race through a parent's mind, and, if you're the new dad, one of those thoughts is: don't drop this. Along with warmth and trepidation, though, comes hope for the child's future, a future full of possibilities. The child may discover a cure for cancer, lead the nation, establish a flourishing enterprise, or discover a new source of energy. While these hopes may overreach the child's capability, that capability has yet to be determined. The parents' hopes have no bounds. Despite the unlimited potential, however, few parents will dream that their child will discover a new technique for meditation or become a free spirit, uninhibited by worries about shelter or money. Interacting with BTBB reality seldom plays much of a role in parental aspirations. Successful interaction with ATB reality, the practical reality that tends to dominate our thoughts, is far more likely. Even if the hope is for the child to write a masterpiece of literature, the target of that hope is society's recognition and the fame that goes with it. The dream is not about some abstract, "useless" tome that nobody reads.

Anxious parents who are overeager for their child to succeed often unwittingly damage a young life. They forget the most important rule of parenting: each child is an individual human being with unique talent, ambition, and potential. Interference from parents, though meant as guidance, is often too narrow—restrictive instead of freeing. It tends to concentrate exclusively on interaction with ATB reality. It ignores the other reality that's just as basic, just as important, and just as meaningful.

Some well-meaning parents work out a complete plan for their children, a plan that may be designed to compensate for their own personal failures. They relive their own lives through a daughter, say, coaxing her along a defined path, warning her about life's pitfalls and insisting on influencing her choices. The unsuspecting youngster may not realize what's happening until it's too late and, when she does, finds herself in a situation she despises. She may have already refused the man she loves. She may have missed that European trip the art teacher sponsored. She may have wanted to take ballet lessons instead of going to computer camp with the neighbor's daughter.

The physical is emphasized in America to such a degree that it's difficult to find the right balance. For example, young males frequently interpret society's tendency to glorify athletic prowess to mean that success in sports is the highest form of achievement. Many a young boy considers himself to be a failure simply because he can't judge a fly ball.

This assertion is not intended to denigrate athletic achievement. Most boys and many girls love sports, and athletics provides them with much more than body-building. It offers peer comradeship and support, as well as the lesson that failure is temporary and success may be just ahead. Also, sporting events aren't purely physical. The mind must be alert, and it must comprehend official rules that can be quite complex. Beyond that, overriding it all, is a valuable trait known as sportsmanship. However, no matter how hard a youngster works or how strong his spirit, if he doesn't have the strength or the coordination, he won't make the cut. The small, awkward boy fails. Even if he finds an outlet where his talents shine, be it in design or song or mathematical skill, the scars left from athletic failure may never completely disappear.

For young girls the issue is often their appearance. They quickly learn that beauty includes figure, facial features, and a personality that brightens the atmosphere no matter how gloomy the weather. A teenage girl understands her status relative to her peers. The natural beauty preens and uses her looks to her advantage. Her classmates fall in line somewhere behind her, all working desperately to present themselves in the best possible way. Although beauty and charm are far more complicated than most teenagers realize, the physical plays just as critical a role for the young female as it does for the young male. Some might say that personality isn't physical, but in this context it is. Here personality means pushing the right buttons to influence the immediate reality. The "social" personality can be quite different than the nonphysical, actual personality that surfaces in a more private setting.

The ultimate criterion for physical success is wealth. Although social status isn't based solely on affluence, wealth plays an important role in determining where a person stands. A teenager with a sports car enjoys a higher status than someone who drives a beater. Financial resources, coupled with the secure upbringing that goes with them, breed popular offspring who learn that success comes from material accomplishment. There may be the oddball Dreamer in the upper-crust crowd, but since most of them are the offspring of Doers, they also tend to be Doers.

As mentioned before, at first there appears to be little cause for concern with a child leaning strongly in the ATB direction. The athletic boy savors high school celebrity, works for and may be awarded an athletic scholarship. However, if he pays minimum attention to academics and uses up four years of college eligibility without absorbing an education, odds are that he will eventually discover that his sports talent isn't sufficient to make it to the professional level. At age twenty-two, the young man is lost and left with few options. He may live the rest of his life trying to eke out an existence, never regaining the glory of his youth. Imagine peaking at age twenty-one.

Another child may find money exciting and rewarding. This child works hard to get a practical education, which may or may not be formal. However the education is acquired, its purpose is to enable the individual to exploit the world, accumulating as much wealth as possible in the process. As a successful adult, wealthy beyond most people's dreams, the tycoon realizes that money doesn't buy love, commitment, or warmth. In the most extreme cases, the despondent soul gives up and commits suicide.

Uninhibited interaction with ATB reality can be as tragic as uninhibited interaction with BTBB reality. Perhaps it's even worse, since exclusive interaction with ATB reality is seldom recognized as a problem until it's too late. And, if a parent is encouraging the interaction as the proclivity grows, who is there to intervene?

At some point, most children turn to their parents for a sense of priority. They look for guidance on how to interact with reality, and that's when parents must fully understand the implications of any advice that they give. When the choice is either financial gain or intellectual achievement and parents make a recommendation based on their own preferences rather than the child's, the result can be harmful. For example, if the choice is based solely on the physical, such as insisting on pursuing an education that will lead to a lucrative profession, a healthy attitude toward the other reality may suffer. This chapter is about extreme Doers, but the same rule applies to Dreamers. If the parents insist on training a child exclusively for the arts, there will be no balance for the Dreamer, and the potential for a fulfilling life as an adult is reduced.

The key to successful childhood development is a proper set of guidelines that are neither too restrictive nor too lenient. That equilibrium is difficult to establish, but in most cases true, unselfish love will dictate when to back off and when to apply pressure. A parent can recognize a child's inclinations better than anyone else and can adjust based on that recognition, agreeing to art courses instead of science, or trade courses instead of college prep, all dependant on the individual.

A Doer mentality is healthy, but it is only healthy if there is a reasonable balance with BTBB reality. Doers have made America strong, and America will stay strong, but only if BTBB reality isn't ignored. If material success becomes the end-all, however, if the physical overwhelms the nonphysical so that it stifles interaction with BTBB reality, fanatic Doers are the result. These extreme Doers take pride in their accomplishments, and, since society offers them so many accolades, there is seldom any force to remind them that ATB reality is only one aspect of life. Thus, the tone is set, and the whole society begins to suffer from a deficiency of BTBB interaction.

Fanatic Doers' insistence on ATB interaction can have detrimental effects on other lives as well as their own. They may encourage ATB interaction in their relationships to such an extent that they stifle other people's personal inclinations. A lifetime of repression may cause friends, spouses, or offspring with a bottled-up Dreamer attitude to suddenly change their outlook on life, resulting in havoc and consternation for those affected. Imagine being married for thirty years, doing everything together, and apparently thoroughly enjoying each other, only to be suddenly confronted with the demand for a divorce. If a spouse decides that he or she has missed living a complete life, this can and does happen. Everyone is shocked. The socially accepted explanation is that the unsettled partner is having a midlife crisis of some sort, but the problem may be caused by a lifetime of ignoring BTBB reality.

Consider this fictional example. Tris had a gorgeous home on a wooded acre, dressed beautifully, and never drove a car more than a year old. All three of her children graduated from Ivy League schools. One daughter became a specialist in oncology, and another advanced to head a Wall Street investment firm. Her son moved to Chicago's Loop and specialized in corporate office design. Her husband, Justin, treated Tris like a queen. What else could she possibly want out of life?

The answer may have been something as simple as a poem that touched her heart. She needed soft, original words to satisfy her dreams. When accused of being insufficient, Justin was far too upset to try anything like poetry. Besides, he had more on his mind than just Tris. The south lot needed a new fork lift, and the west lot needed more under-roof storage. He'd worked his butt off since graduating from college, buying a one-man lumber supply firm and turning it into a regional power-house. His will power drove the growth. It took 80-hour work weeks and many vacation-less years to pull it off, but Justin had succeeded. And during that time he never missed his boy's ballgames—not even once—and he gave that kid and his daughters all they could ever want, including a little cash infusion when they got their bachelor's degrees. He treated Tris to a dinner at the Evergreen at least once a week.

Then the marriage exploded. Surprise!

But the outcome needn't have been a surprise at all. If both partners had been observant, they would have noticed that something was missing. Sunday church service might have been enough to help. Going to a play once in a while. Music. A good laugh. There were plenty of opportunities missed, and that's the tragedy of Doer fanaticism. To ignore BTBB reality is to ignore a major part of total reality, and eventually that omission will bite.

A commonly prescribed cure for an unbalanced life is to pursue a "well-rounded" lifestyle. The issue here, however, is more fundamental than that bland expression implies. People tend to think of "well-rounded" as meaning well-versed in a variety of human activities, so that they can tackle all sorts of problems. A well-rounded plumber knows enough about electricity to keep from making a catastrophic error when installing an electric water heater. A husband can do household chores without scratching the woodwork or staining the carpet. A wife understands the workings of a car well enough to get the oil changed and check the tire pressure.

However, all of these "well-rounded" traits involve interaction with ATB reality. They help people survive in the physical environment but do nothing to encourage them to interact with BTBB reality. A **real** well-rounded education provides insight into total reality, a reality that includes both the physical and the abstract. A **real** well-rounded lifestyle involves interacting with both ATB

and BTBB reality. It involves combining the contemplative and the reflective with the practical and the immediate.

We've all known Doers like Justin. They are fanatic Doers, who become that way by succumbing to the pressure to become a "success." They become so engrossed in their choice of reality that they don't realize they have a problem until their life literally disintegrates. These fanatic Doers are usually easy to identify but difficult to confront. Solving their problem is challenging, since the Doer is only performing normal tasks at what the Doer believes to be a higher level. As pointed out in the previous chapter—try taking the Doer to a live comedy—the fanatic Doer needs to be exposed to BTBB reality in order to learn its value.

Radical Doers who don't learn to appreciate BTBB reality can go beyond the fanatic stage. The most extreme type is a Manic Doer, a person who violates accepted Doer reality standards and thus becomes society's enemy. The Manic Doer's abnormality is usually more difficult to expose than a Pseudo Dreamer's, since a Manic Doer may be a leader of society and have plenty of money available for influencing others. Manic Doers, like Pseudo Dreamers, distort reality to such an extent that the reality they address isn't reality at all. It is a phony reality invented to conform to their extremist views. In the next chapter we'll delve into the realm of Manic Doers.

Chapter 29
Manic Doers and Society's Response

What are Manic Doers? Does society take notice? If so, what is its reaction? After defining Manic Doers, we'll see that society not only recognizes them, but that its reaction forms the basis of criminal law. We'll analyze that reaction in depth when considering the most outrageous Manic Doer act—the taking of a human life. Finally, we'll see how authorities can go overboard and, armed with experience in regulating Manic Doers, try to regulate things they shouldn't.

We've defined a Doer as an individual who emphasizes interaction with ATB reality. A fanatic or extreme Doer is one who places so much emphasis on the ATB world that virtually all interaction with BTBB reality is omitted. Song, art, abstract thought, and meditation are not part of the extreme Doer's life, unless there's a way to make a profit from them or to gain prestige that can be leveraged into profit.

The Manic Doer takes the extreme Doer's attitude to the next level, the level where the only rule is not to get caught. That is, the Manic Doer will do anything to improve interaction with ATB reality, and will do so at anyone else's expense. Clever Manic Doers will find a narrow line where they can practice their connivances without paying the consequences. Avoiding punishment is important, because they value their freedom, as well as all the other

226

physical pleasures that ATB interaction allows. Getting caught is the equivalent of failure. This wariness doesn't mean that Manic Doers shy away from intrinsic evil. Instead, they either practice schemes where there is no law, or work along its edges where the lines of control are blurry enough to conceal their activities. As far as others getting hurt? The Manic Doer couldn't care less.

To defend itself from these unscrupulous individuals, society has built an elaborate system of criminal law. Violence is prohibited, other than to protect society as a whole. There are elaborate accounting rules to control financial statements and assure honest reporting. There are statutes governing the environment, transportation, commerce, and construction. These laws are calculated to protect the weak from the strong, the victim from the predator. But although society works hard to rein in Manic Doer activities, its laws are never perfect. In addition, technology and culture continue to evolve, promoting an unending procession of creative new schemes. Meanwhile, honest, balanced Doers plod along, scrupulously obeying all the laws, while the Manic Doers grab whatever is available for grabbing, continually changing tactics to avoid prosecution.

Society uses all the tools it can muster to protect itself from Manic Doers. Anyone who breaks the rules is punished in a clearly prescribed manner. This clarity is critical. Criminals are warned of the consequences of any crime, and broadcasting the risks deters rational people from engaging in illegal activities. To punish those who are caught, prisons are built and fines are imposed, anything to discourage disobedience of the rules. At the same time, America's law schools churn out graduates like popcorn. Some of these graduates enter politics, planning to author even more rules to assure a fairer interaction with ATB reality. Others become judges, presiding over courts and making pronouncements defining proper ethical standards. All of this because interaction with ATB reality **must** be controlled. If it isn't, Manic Doers will dominate, and there will be chaos.

Unfortunately, establishing fair and effective rules can be challenging. First, there is a fundamental dilemma between rewarding productive Doers and penalizing Manic Doers. Society is anxious to help those who benefit it the most, and frequently those members are extreme Doers. Knowing that, Manic Doers conceal their illegal activities by masquerading as Doers engaged in honest enterprise, the very ones the public wants to encourage. Thus, society finds itself torn between assisting honest Doers and protecting itself from Manic ones.

Politics mirrors this division. Some influential politicians and community leaders promote tax incentives and reduced regulation to stimulate development. Others take the opposite approach. They advocate increased taxes and stricter controls over business operations to prevent Manic Doers from making "windfall" profits or taking advantage of lax oversight. Thus, society runs the risk of hurting honest and productive enterprises in the attempt to thwart the machinations of those who take advantage of the public for their own ends.

Second, defining the problem that a law is meant to address and making those laws clear can be difficult. Perhaps the best way to understand this issue is to examine society's rules regarding the greatest insult in interaction with ATB reality—the taking of a human life.

Manic Doers value just one life—their own—and one would think that effective governments would have clear, firm, and fair policies concerning life and death. In many cases they do, but, even in the twenty-first century, areas of ambiguity remain. For example, when does life begin? Face it, the lack of a clear definition of that point opens the door for debate. Even within a given culture, individuals have widely differing views, as the abortion debate plainly demonstrates. Do Manic Doers play a role in this controversy?

Most abortion proponents are sincere, well-meaning, intelligent people who, in proclaiming a woman's right to choose, permit the destruction of unwanted fetuses. Yet their opponents condemn that procedure as an act of murder. The purpose of these advocates, however, is to assist a woman who doesn't want to bear a child. That's not Manic Doer motivation. To some, it may be misled Doer activity, but the goal is to help a fellow human being, not to gain material wealth. On the other hand, if one believes that such action is murder, then the woman who insists on the abortion could be considered a Manic Doer. No wonder this ATB reality issue is so controversial. If abortion is murder, then it is Manic Doer behavior that must be controlled. However, if abortion is merely allowing a woman to control her own body, it is normal Doer behavior that requires little oversight. This dilemma highlights society's difficulty in defining a life-and-death issue.

Euthanasia is another area which raises many of the same questions and problems as abortion. Watching a loved one in severe pain can be an agonizing experience. Even if painkilling medication is available, when there is absolutely no hope of survival, the question of allowing the person to die often comes up. If an individual has no consciousness and no possibility of regaining consciousness, what is the point of continuing that life? These are difficult questions, and there is no clear answer, no matter how objective or even religious a person may be. Circumstances vary. While one medical opinion may say that there's no hope, another could differ. Also, even if all the medical experts concur, there is always the possibility of an exception.

Society must treat the issue of terminally ill patients with extreme caution. Any expedient or simplistic solution provides an opportunity for Manic Doers. They use such opportunities for grabbing money or power or both. They have few scruples and are primarily concerned with improving their own ATB circumstances. Given the chance, they will take a simplistic solution and run with it, whether it be that life must be preserved no matter what the circumstance, or that life can be terminated once a set of conditions is met.

Abortion and euthanasia show that establishing clear rules about life and death isn't always straightforward. And, if society has difficulty with such a basic question, think of the struggles it must face with more subtle ones. As a result, society often unwittingly helps Manic Doers even as it attempts to thwart them.

If left to themselves, Manic Doers will consume society's resources and destroy the very fabric of society itself. Not having clear rules invites catastrophe. If the laws aren't clear for construction, structures will collapse. If they aren't clear for financial systems, economies can fail. If they aren't clear for the environment, human existence itself is threatened. Manic Doers will exploit any ambiguity to the fullest. They have no conscience and only one priority—themselves. This problem gets even worse when the issue is between societies, that is, when there is war.

The taking of life, including innocent life, is an integral part of waging war. Yet for centuries warriors have been lauded as heroes, and war, whether offensive or defensive, was deemed a noble activity. The horror and carnage of World Wars I and II finally convinced most of Europe that avoiding war should be a primary goal for all nations. Arguably World War I should have been sufficient to send home that lesson, but not everyone was convinced. Some still held onto the attitude that a glorious military victory would establish their superiority over others. Hitler's diatribes exploited that belief, but the catastrophic results of his actions have finally persuaded most people that such a belief is wrong. However, even after World War II, the Communist system went as far as it dared, using force to extend its influence in Europe and Asia.

In the twenty-first century most societies, though they still honor their warriors, are repulsed by violent conflict and agree that only defensive wars are acceptable. The use of military force is permissible if attacked, but only if attacked. The logic is valid, but defining what constitutes an attack is not always clear-cut. A national leader who believes that the culture is being threatened by some external force, may decide to attack that force. Once under attack, the victim will respond, firmly asserting that the response is entirely defensive in nature. Thus, two societies feel obligated to try to destroy each other. Both claim that their actions are legitimate, because they are acting in self-defense. Even though the combatants may sincerely believe they are only protecting themselves, Manic Doer conduct takes over. To conduct war, participants must inevitably use Manic Doer tactics—if they want to win.

Of course, Manic Doers themselves are usually at the center of hostilities. Criminal leaders, who are unquestionably Manic Doers, will initiate and intensify belligerent actions if their aggression is not challenged. The world appears to understand this and has organized itself to respond in a variety of ways. If it does so overwhelmingly, the criminal can often be controlled without military action. A united front eliminates the opportunity for the aggressor to claim legitimacy, since the "enemy" would be the rest of the world. Unfortunately, uniting the nations of the world with their differing priorities is often impossible. Thus, world organizations fail, and wars continue.

This discussion demonstrates how difficult it is both to define and to establish satisfactory rules to regulate Manic Doer transgressions. Society has so far been unable to successfully control even the most severe forms of Doer mania. Nevertheless it must continually work to improve its capacity to do so.

Since Manic Doers can morph into ruthless aggressors who destroy fundamental institutions, society naturally controls and punishes their behavior. On the other hand, a healthy government virtually ignores most Dreamer issues. The reason is simple. Why bother? Although Pseudo Dreamers may be unproductive, they are usually harmless and require little oversight. Any risk from a Pseudo Dreamer is minimal compared to the chaos Manic Doers can provoke.

But while regulating Manic Doer behavior presents more than enough problems to keep most governments busy, some don't stop there. For them, Dreamers need to be monitored as well. Thus, they establish rules that regulate both ATB and BTBB behavior, blending the two realities and attempting to control both. Effective, helpful governments only control aberrations of interaction with ATB reality with the goal of assuring a safe, healthy environment for all while keeping a watchful eye for Manic Doer threats. Ineffective, unhelpful governments seek to control things that don't need controlling—such as art—or to control things that can't be controlled—such as thought—or to control things that shouldn't be controlled—such as religious belief. Some of the most nonfunctional governments are those headed by religious zealots who believe that their faith requires them to establish rules governing all aspects of life. Questioning these leaders is labeled as heresy, and violators are condemned in this world and damned in the next, leaving them no hope. This type of perverse government has been all too common throughout history.

When a government attempts to regulate a BTBB interaction such as artistic freedom, it often succeeds in encouraging the very ideas it wants to quash. For example, a government might support artistic endeavors financially, but then attach strings to the money. Sometimes those strings also prohibit specific types of creative production, and the penalties for breaking the rules can go beyond simply withdrawing monetary support. In response, artists who want to go their own way can either refuse the money or take it and ignore the prohibitions. Either way, they recognize the strings for what they are and try to outdo each other in violating the rules, in secret if necessary. History has demonstrated that attempts at artistic censorship are seldom effective. Dictatorial societies may try, but their efforts only suppress artistic candidness; they don't eliminate it.

The concept of separation of church and state is another example which shows that governments are better off when they don't regulate beyond ATB reality. In the twenty-first century, most of the Western world accepts this principle. Yet despite its Christian foundation, it took almost two millenniums for the West to practice what the Bible's New Testament says so simply: "Render to Caesar the things that are Caesar's and to God the things that are God's." The reason it took so long is that the existence of two realities isn't set in human consciousness. As long as the "one-reality" concept is embedded so firmly, some governments will be tempted to regulate it all, including the proper way to worship. Doing so led to generations of warfare in the sixteenth and seventeenth centuries after the Protestant churches broke away from Roman Catholicism, and governments tried to force their subjects to accept one competing faith or another. As a result, hundreds of thousands were killed, and much of Europe was laid waste. Unfortunately, history is replete with such examples. On the other hand, governments that concentrate only on ATB regulation allow their citizens the freedom to think and believe as they choose, thus encouraging both the diversity and the creativity that will help a nation to flourish.

A healthy society will provide sufficient controls to ensure that Manic Doers are kept in check despite social, economic, or scientific change. However, any attempt to extend its influence beyond that realm can lead to undesirable, even dangerous, complications.

Thus our discussion of the most extreme type of Doers is complete. In the last few chapters, we've looked at individuals all the way from Pseudo Dreamers to Manic Doers. Few would argue with the claim that those at either end of the spectrum are extreme cases, individuals who have failed to achieve a sense of balance. Let's now turn our attention to procuring a proper balance, something we all pine for in our lives. Up until now, however, we weren't quite sure what we were balancing. We knew that both the sublime and the practical existed, but weren't sure exactly why. Now that we know why, finding that balance should be easier. In the next chapter we'll pursue a method to help us discover a balance between interaction with ATB and BTBB reality.

Chapter 30
Finding a Balance

It's natural for people to follow their inclinations. If we're inclined to eat or drink, or pester and pursue the opposite sex, or make money, or exercise, or garden, we do. And, if we're reasonably good at whatever our tendency is, we'll do it more frequently. Most of us are more proficient at ATB reality, since that is what assures our survival. If there's no survival, BTBB reality doesn't matter. In addition, backing away from ATB reality can make our lives more uncomfortable. The result may be a smaller income and the inability to provide all the things we want for ourselves and our families. So most of us plow ahead, working desperately to succeed in ATB reality, concentrating on it to such an extent that interaction with BTBB reality is forced into the background. The ability to balance the interactions, so that the appetite for both is addressed satisfactorily, is what we need to strive for.

While society tends to promote ATB activities, most people are left on their own to understand and incorporate BTBB reality into their lives. There may be help from well-meaning teachers, who attempt to expose their reluctant students to the "finer points of life," but few youngsters appreciate the effort. Instead, they seek the physical gratification that comes with success in the ATB world. There are far more enticements and incentives in that direction than in the subtle and often unnoticed rewards associated with

BTBB reality. Social pressure and the instinct for survival nurture the obvious, and ATB interaction is the result.

This natural propensity can lead to an overemphasis on ATB reality. Occasionally, those who recognize this phenomenon may decide to swing their personal pendulum in the opposite direction, toward BTBB reality, but that swing can also go overboard. Thus, overcompensating for a perceived deficit is one way to cause a problem at either end of the spectrum. We've already seen the results in Pseudo Dreamers and Manic Doers.

How can a person avoid these imbalances? First, there must be a recognition that each individual is unique. There is a proper fulcrum point where interaction with BTBB reality balances interaction with ATB reality, but its location is different for everyone. We will call that proper placement the "balance point." Some are more comfortable with ATB interaction, so the emphasis must be in that direction. The stronger the proclivity is, the further the fulcrum point must be placed toward the Doer end of the spectrum. The placement is independent of the Doer's talent. That is, even if a Doer is having difficulty earning a living, the fulcrum point shouldn't change. Perhaps the career choice should be modified, but it should still emphasize the practical. Similarly, a Dreamer must place the fulcrum at the Dreamer end of the spectrum. That allows the Dreamer more opportunity to consider the impractical and tinker with the fanciful.

Sadly, the Dreamer spectrum offers fewer career opportunities, so Dreamers often find their activities forcibly skewed toward ATB reality in order to make a living. They have to use their free time to compensate. Weekends and vacations are spent enhancing their "hobby." This predicament tends to warp the Dreamer's view of interaction with ATB reality. The Dreamer resents ATB's overbearing clout as well as the Doers who flaunt their success with it. The resentment may grow, and with it an imbalance toward BTBB reality.

Though the Dreamer must satisfy the desire for BTBB interaction, the inclination must be kept under control. A Dreamer needs to fight off any ill feelings, as well as find a way to survive. Earning a living may be tiresome, but it is necessary. Instead of resenting that necessity, Dreamers must somehow force-fit BTBB reality into their lifestyle. Fortunately, it's old-hat for some, who come home exhausted from work and happily immerse themselves in mastering a musical instrument or composing poetry. Eric Hoffer earned his everyday existence on the West Coast piers, but he spent his free time contemplating philosophy. Later, when his talent was recognized to the extent that he no longer needed to depend on a paycheck, he continued to work on the piers. He decided that the work grounded him and gave him a balanced outlook on life.

Although Dreamers are more likely to struggle in finding the proper niche for their operating zone, some Doers have similar problems. Doers whose parents push them into "artsy" activities and don't get sufficient opportunity to pursue their own inclinations may become stuck as Dreamers. Their frustrated Dreamer parents insist that they pursue BTBB interaction and force music lessons on offspring who yearn for the ball field and teammates. Parents drag them to museums and libraries instead of letting them go swimming or skateboarding.

Combating forces that oppose a person's natural inclination makes life difficult, but it can be even worse if individuals don't know where they fit in the Doer/Dreamer spectrum or aren't sure of their motivation. Dreamers are notoriously insecure. Do they really yearn for the contemplative, or are they simply afraid of competition? Is that why they back off any challenge? On the other hand, while Doers are usually practical enough to be confident, they too must guard against false attempts at achieving balance. Is going to the opera providing them with BTBB interaction, or are they simply trying to improve their social status by showing up at an artistic event?

Knowing one's real preference is critical, just as critical as recognizing an unhealthy imbalance or an illegitimate interaction with either ATB or BTBB reality. So, many readers will ask, shouldn't there be a test to find out? Books like this are supposed to provide some sort of test, aren't they?

> Okay, let's take one. Answer the following questions with a number 1 through 5: 5—strongly agree;

4—agree; 3—neutral; 2—disagree; and 1—strongly disagree.

1. I find the eight-hour workday too short.
2. I love to shop for a bargain.
3. I hate to wait in lines.
4. I love to read fiction, the more fanciful, the better.
5. If it's a warm summer day, I rush home to sit under the oak tree in the backyard.

To score, add your answers to questions 1 through 3 and subtract the answers to questions 4 and 5. If your total is more than 3 you're a Doer. If it's less, you're a Dreamer. Oh, and if you really wanted to take the test, give yourself a five-point bonus. Taking tests like this, that is, searching for a quantitative answer to a complicated question, signals Doer predominance. Most Americans love these tests.

Going any further in attempting to define yourself is risky, since a person can have different balance points, depending on the situation. An example would be a hard-driving Doer who's devoutly religious. Extreme Doers often find that religion tempers their lives and gives them balance. They know that they need relief from the hard-hitting reality they constantly face, so these Doers temporarily become Dreamers. On the day of worship, their "balance point" is toward the Dreamer end of the spectrum. However, when the workweek begins, the pursuit of their career commences, and these Doers return to being their old selves, perhaps even Manic Doers. Similarly, the artist who despises the money-grabbing tactics of capitalists might be a fanatic on the soccer field on Sunday afternoons. For some Dreamers, sport cleanses the spirit and revives the soul. Fishing, hunting—even war!—did it for Hemingway.

Finding out how to balance interaction with the two realities isn't easy. Both the method and the amount must be addressed. To be successful, the individual must establish an attitude rather than a fixed position. That is, to insist on playing the piano for an hour every day, no matter what, may not only be impractical, but could be harmful. The same rule applies to the opposite personality type. Just as the ardent Dreamer must allow for the practical, the Doer must allow for the nonmaterial.

So how is the proper attitude established? First, before any imbalance can be addressed, it must be recognized. Let's find a method to do just that.

A busy life often conceals the issue, and many who lean too far in one direction or the other believe that they're doing just fine. If there is a problem, it's someone else's. All of life's variables and complexities easily mask both the imbalance and its cause. Nevertheless, we'll find a way to look for one, and we'll find that method in an odd place—a factory floor.

If there's any place where ATB reality must be addressed and done so immediately, it's in a production facility. Production depends on solving problems on the fly, using any method or strategy that the factory worker can invent. There are no rules in factory work other than to make the product correctly and ship it in a timely fashion. Whether working on the production line or in support of that line, factory workers understand the consequences of poor management, shoddy engineering, and backstabbing politics. These same workers also realize that if the factory is operating properly, the managers, engineers, and line workers from the least skilled to the trained specialist will cooperate with each other to solve any problem that surfaces. The issues can be far-ranging: from parts out of spec, to a new government regulation, to a competitor introducing a product that threatens to make the factory's product obsolete. A sloppy coworker may arrive with a hangover and ruin a shift's worth of work, or the management may introduce a new quality scheme that no one understands. A delivery could be late, or a million other things could go wrong. If it can happen, it will happen on the factory floor.

This makes the factory a perfect place for Doers. Doers react as only Doers can and force production on, hurdling all the obstacles along the way. The result? The product ships and customers can give immediate feedback on the timeliness, quality, and performance of the shipment. Doers love that immediacy, even in cases where there's a problem and the feedback isn't flattering.

Manufacturing plants teach their workers many lessons, but the most important one is that to solve a problem, it first has to be identified. A method for that identification is the crux of the reason we're discussing factory production.

In a factory, finding the fundamental reason for a problem is often more difficult than solving the problem itself. The symptoms may be obvious, but the root cause can be hidden under layers of politics, misunderstandings, falsehoods, and prejudice. The Japanese found a way to expose the cause of a problem, and their discovery turned out to be—as great discoveries generally are— quite simple. All one needed to do was to take away inventory, that is, to limit the amount of product within the factory to its bare minimum. Once all the extra parts were gone from the factory floor, no problem could be hidden. The source of the difficulty was immediately exposed, naked for all to see. Once the root cause was addressed, ambitious Doers could go to work on a solution. The barely-out-of-spec part was either kept out of production, which stopped the whole line and focused on why the supplier couldn't meet the specification, or, if the specification was too tight, it was loosened. Either way, only parts that caused no difficulty were allowed to be used.

The Just-In-Time philosophy caught on in America and has forced competitors to either adopt it or go out of business. Although factories are Doer domains, and Just-In-Time helps these Doers, the main point of this lengthy illustration applies to both Doers and Dreamers. **The point is that we need to define the problem before we can solve it, and that rule holds true in developing a proper attitude for interacting with both realities. We must strip away all of the peripherals that society uses to "explain" human behavior and concentrate only on the imbalance.** Once we clearly focus on the out-of-balance situation between ATB and BTBB reality, we can find out why it exists and take steps to resolve it. That is, once we understand which aspect of reality is being overemphasized, we can work on ways to increase interaction with the other.

So now we have our method. We must strip away anything extraneous and then take a close look at our life. Let's first review our status with ATB reality. Again, ATB reality takes priority. We can't exist if we don't address it. Therefore, the initial step in any search for imbalance is to strip away all other interaction and isolate that portion of our lives that deals solely with ATB interaction.

Life is complex, easily as complex as any factory operation. Social pressures, ambition, family considerations, and a myriad of other forces continue to bombard us. Earning a living, raising a family, and dealing with others are all embedded in ATB reality. To isolate that interaction means to put aside any BTBB activity and concentrate only on that which allows us to physically survive. Forget about finding the perfect color for the living room walls and think about patching the cracks and fixing the light fixture. Stop pondering over which outfit best suits your mood at the moment and concentrate on getting the soiled clothes to the cleaners. Put aside all the BTBB in life and concentrate on the ATB.

We'll use Ted for an example.

Ted is about as average as they come. He's quite satisfied with himself. He has a great family, a wife, and three rambunctious daughters ranging in age from four to ten. His accounting job pays well enough to support a three-bedroom home in a good neighborhood. "And this car is only a year old," he says, proudly showing off something called a crossover. "I get a new car every three years." His smile broadens. As far as ATB reality goes, Ted figures he's doing better than average. And, if there's any imbalance in his life, he certainly isn't aware of it. Nevertheless, he agrees to take a look.

At first he's a little confused about what to do. Once he understands that the exercise is only mental and that he only needs to imagine removing those activities that don't directly help him survive, he's ready. After some thought, he takes TV out of his life. But then he puts it back in, because he figures that watching the news is ATB reality. He goes to church every once in a while, so he eliminates that activity. He doesn't read much, but he removes that too. And the TV that's not news or sports—that goes. Sports isn't critical for survival, of course, but there's nothing particularly cerebral about it, so it could stay, right? On the other hand, just watching sports offers no exercise. No, he decides, sports on TV goes. He never meditates, so that's not an issue. After thinking a little more, he decides that's about it. The rest of his life is ATB.

"I'm a Doer," he proudly announces. "That'll make the boss happy." He laughs and heads for the garage. There's a tree branch that's too low and gets in the way when he's cutting the grass. He's been wanting to cut that branch off all summer, and now that he's found his "balance point" he's going to get it done.

But wait. Ted isn't finished. He may have stripped the BTBB away, but that isn't the end of the exercise. He still has to make an analysis of what's left.

"Most of my life is left," he says with a smile. "Unless I'm supposed to get rid of the kids!" He laughs and marches for the low branch.

Some prodding makes him stop. Saw still in hand, he sits down on the grass and thinks. He's sure he's a Doer, but now the question is, how extreme a Doer is he? He mentions that he once turned down a promotion, because it meant moving to Texas—too far from his family. "That's Dreamer, right?" he asks.

Looking back on his life, he begins to reminisce about high school and acting in the senior play. The memory stirs up the long-forgotten joys of appearing on stage. The more Ted thinks about it, the more antsy he becomes. Getting up, he fingers the saw handle and complains that he simply doesn't have enough time for stuff like that. He can't even get that branch cut!

Then a secret spills from his lips. Ted had always wanted to write. When he was in high school, he even wrote a different version of the play he was in, making his character a villain instead of merely the "Second Policeman." He never showed the script to anyone. His eyes light up. "I've still got that. It's in the attic." He hurries away and comes back thirty minutes later waving a folder.

There was some Dreamer in Ted, and the more he thought about it, the more he realized he had suppressed its influence. His family and job had kept him too busy. "Maybe one day…" His voice drifted off.

By stripping away all BTBB interaction, Ted realized that he didn't have to drop very much. What he cut out was so minimal, in fact, that he realized that any Dreamer tendencies he harbored had hardly any time to get out.

Once BTBB interaction is eliminated from a lifestyle, we can make a judgment. If the physical side of our life is suffering in a recognizable manner, we're addressing ATB reality insufficiently or ineffectively. Whether we're lax in personal hygiene, losing weight, missing important appointments, or constantly fighting off creditors, the cause is inadequate interaction with ATB reality, and that situation needs to be addressed. On the other hand, if we're like Ted and overemphasizing ATB reality, we may find that we have few friends outside of our professional life, or we may notice that we have an obsession for accumulating money, or we may feel burned-out. Or we may discover that we have suppressed a basic longing. Perhaps when that desire eventually surfaces, the result will merely be regret. However, the consequences may be more serious and emerge as they did for Tris a few chapters back, a woman who was so unhappy with her situation that she left her husband. Ted probably should spend a little less time on his profession, so he can have more free time. Perhaps he wouldn't be able to afford a new car every three years, but he would have time to pursue his dreams. When we discover we're overemphasizing interaction with ATB reality, we must shift some attention to BTBB reality in order to enjoy a balanced, fulfilling life.

Notice that once we have our interaction with ATB reality under control, the need for more or less BTBB interaction will become obvious. Again, the key is to isolate ATB interactions so that extraneous interactions will not contaminate our judgment.

After making a decision about our balance point, we may have misgivings about the appropriateness of its location. A parent or spouse may have convinced us that our fulcrum should be placed differently than it is. Such external influences often instill misleading guilt. Seeds may have been planted in childhood that continue to tell us we're not doing what we should. However, these seeds are simply more distractions and "excess inventory" that hinder us from making a good judgment. Again, strip these away!

We need to continually remind ourselves that both realities are equivalent. There's no reason to feel guilty if we have an inclination away from the arts and toward making money. An artist has no grounds for berating the entrepreneur who has made a fortune in fast food. On the other hand, craving contemplation and avoiding structure isn't bad either. Both inclinations must be taken into consideration to have a full and successful life, but they both must be kept in balance. Problems will occur if one of the inclinations is either ignored or pursued to the extreme.

This need for balance further demonstrates the necessity for nurturing our youth properly. The failure to do so can sow guilt and tempt a child to interact with reality unnaturally later on. If a child isn't interested in piano lessons, leave that child be. If a child wants to perform manual labor, let that child go ahead. Plumbers often earn more than the so-called educated elite. Actually, that child may eventually develop into a Dreamer, and a career like plumbing can pay enough to allow for a pursuit of music, art, or writing. Being true to one's nature is what counts, and that's what should be emphasized to young people, occasionally reminding them that our nature includes interaction with both ATB **and** BTBB realities.

It's essential to be honest with ourselves and respect our own "balance point" as much as possible, realizing that circumstances will always influence us to some degree. Marriage, for example, involves a lifetime of compromise, so finding a viewpoint on reality that's acceptable to both partners requires give and take. A change at work can necessitate accommodation as well. A promotion may require a person to engage in either more or less theoretical effort (i.e. more or less BTBB interaction). The important thing is being aware of our "balance point" and working to keep our lives as close to it as possible.

Again, there's nothing new in making such a search. Long before Einstein revealed his Special Theory of Relativity, people were looking for ways to balance the practical and the sublime, making use of mysticism, psychology, and many other tools in the attempt. Until now that effort has been full of second guessing and frustration. In this treatise we've taken away that "excess inventory" as well. By defining and analyzing total reality, we are now able to understand the reasons for the forces that lead us in one direction or another. This knowledge will allow us to decide what works for ourselves, so we can proceed without guilt or prejudice. Just as there is no reason to condemn someone else for emphasizing interaction with either reality, there is no reason to force ourselves into viewing reality differently than our personal inclination indicates. The only requirement is that we accept the fact that there are two realities, and that both are equal and fundamental.

Chapter 31
Dual Reality—Past and Present

There's nothing revolutionary about referring to a duality in human nature. Literature and the arts have explored this phenomenon for centuries. In this chapter, we'll look into some of the historical explanations of the two realities and see how society, particularly in the West, has depicted them. Then, we'll look at a contemporary view and see how dividing total reality into ATB and BTBB helps us understand human tragedy. Finally, we'll introduce the relationship of BTBB reality to spirituality. We'll expand on this last topic in the next chapter.

Virtually every civilization has conceded that there is more to life than the physical, that, in a sense, there are two realities: the one our senses perceive and some other reality which has been labeled intellectual, artistic, spiritual, or mystical. Explanations as to why the two realities exist have often required some beyond-the-normal capability or a leap of religious faith to fully comprehend them. Even those who question the validity of religion, accept some form of abstract reality.

Most civilizations view the two realities as encompassing the higher and lower aspects of life, which is why the arts and other abstract thought processes are given a special place in society. Recognized talent in writing, music, dance, painting, and sculpture allows artists to achieve high status, even though the unrecognized may be labeled impractical and/or lazy. Doers reward artists who please them, and artists gratefully accept that reward. Providing cash gives Doers a sense of participation, and it gives artists the freedom to practice their art.

Of course, money isn't the real measure of artistic value. Imagine trying to put a price tag on Michelangelo's *David* or da Vinci's *Mona Lisa*. These are masterpieces that no amount of money can buy. Their images capture life, **all** of life, which we now know as interaction with both ATB **and** BTBB reality. *David* and *Mona Lisa* blend the nonmaterial and the physical in unique and beautiful ways. Mona Lisa's smile is erotic yet peaceful. She graces a majestic landscape, balancing visual and spiritual beauty. David is the masculine image of that same dual reality, his expression combining the determination and concern of a thoughtful warrior, while the muscles in his arms and legs shout out his strength. *Mona Lisa* is the Doer **dreaming** and *David* is the Dreamer **doing**. These works of art portray the key to a rich and fulfilling life—a Doer dreaming and a Dreamer doing.

Another example of this duality can be found in literature. Charles Dickens' depiction of Ebenezer Scrooge exposes a Doer who has lost his sense of BTBB reality, and the story demonstrates the consequences of ignoring that reality. Dickens, and the nineteenth-century society he wrote about, realized that those who care only about accumulating wealth become slaves to a world that is cold and unforgiving. People who enjoy the warmth of dance, companionship, and laughter are far happier than misers who spend their days scratching for ever more money. The moral of *A Christmas Carol* shows that life must incorporate that which is timeless, nonmaterial, and universal.

Dickens created a Doer who ignored interaction with BTBB reality, but he could just as easily have chosen a Dreamer turned away from ATB reality. If his emphasis had been on a Cratchit who enjoyed partying so much that he didn't provide for his family's well-being, the results would have been equally tragic. In that version of the story, the plot would have been reversed. Instead of showing an unhealthy inclination toward the physical, it would have shown an unhealthy inclination toward the nonphysical. The ending would have Cratchit going to work on Christmas Day to solve a crisis. The readers would have rejoiced just as much as they had with the original, having witnessed Cratchit give up a chance to enjoy himself in order to turn his life around by saving the business.

Although art and literature have played a significant role in defining and dealing with the two realities, arguably they're amateurs when compared to the activities of organized religion. Religion separates the physical world of everyday life from the spiritual world of a creator, however that creator may be defined. In every religion there are two realities, the material and the nonmaterial.

In some cases the spiritual world has included anything that human society couldn't comprehend. These faiths turned to the mysterious and/or the magical to provide an explanation. After all, what alternative was there? Later, when people understood the physical side better, the supernatural aspect vanished. For example, not too many years ago there were cultures that worshiped the sun. While these societies didn't understand how the universe worked, they did understand how critical the sun was, so it became their deity. Now, of course, the sun isn't considered mysterious at all. It's no longer divine and no longer has supernatural powers. This change in attitude might be equally applicable to natural phenomena that we currently think of as supernatural but which, in fact, we don't have sufficient knowledge to understand.

Having a factual explanation always helps human understanding, and understanding BTBB interaction helps us explain life. Without BTBB reality, much of human experience appears mysterious—as television would have been in an earlier century. Think of what someone in the fifteenth century would have thought when confronted with television. The astonished individual would, at the very least, assign the phenomenon to the paranormal. Since there was no other reasonable explanation at the time, the poor soul would have had no option. Today we are aware of the airborne transmission of electromagnetic waves and realize that there is nothing extraordinary about them. If we didn't have that information, however, we would have to agree with the fifteenth-century assessment. Similarly, life without BTBB reality is mysterious; with it, it's understandable.

Eventually we may understand how all reality works, and there will be fewer phenomena to label as supernatural. In fact, by looking at ATB and BTBB interactions we now understand **why** our reality has a dual nature. The question then becomes more fundamental—why does this duality exist? Both the religious and the nonreligious can work on that question, because both groups have the same craving, the same need for the nonmateriality, timelessness, and universality that interaction with BTBB reality offers. Interaction with both realities occurs in every person's life. It's a mistake to totally ignore BTBB reality, no matter how strong the impulse may be to disregard that which is not based on everyday, practical issues.

Even when both realities are addressed in a reasonable manner, however, life still presents us with success and failure, goodness and evil, joy and tragedy. The secret of life is to use both realities in the proper manner to minimize life's negatives and maximize its positives.

Consider a tragic loss, such as the death of a child. Think of the pain and suffering that the parents and loved ones experience. Such a catastrophe is so traumatic that it can't be fully comprehended without experiencing it. Nevertheless, let's look at that tragedy from the viewpoints of both ATB and BTBB reality and see how those perspectives can ease—if only minimally—the pain.

The death itself is ATB reality. We live in ATB reality. We're embedded in it. Therefore, to prevent a similar tragedy, we must address ATB reality issues. If the death was due to an illness, a cure needs to be found. If it was due to an accident, the cause of the accident must be identified and eliminated. Although a cure may seem impossible to achieve, or the cause impractical to address, an effort must be made. The parents may lead the way. Government or community leaders may lead the way. But to simply accept the death and do nothing is repulsive to most people.

Working on a program to prevent a similar death may help the parents and loved ones in their grief, but it won't take away the enormous loss they've experienced. Nothing will bring the child back, the only resolution which would fully satisfy the parents. However, keeping busy helps; that's the ATB viewpoint. Now let's turn to the BTBB perspective.

Although we don't live in BTBB reality, we interact with it in many ways. Therefore, we can meditate, contemplate, and rejoice in the memory of a life, even though tragedy has cut that life short. And while it was too brief by ATB standards, in BTBB's timeframe that life was long. Remember, BTBB's eternity is our instant! Of course, the child will never experience a full life with all of its potential. Maturity and old age will never occur. But think of the worries and strife that will be avoided as well. What counts is the life that was there for as long as it was. ATB reality somehow cut that life short, and the hurt will never go away. Nevertheless, concentrating on the life as it was can be a source of comfort and help ease the pain. Although the ATB reality possibilities— the cure or the elimination of the cause—may be remote and unlikely, memories of all the laughs, hugs, and joyful events can be immediate, vivid, and shared.

The problem is that human BTBB interaction is not perfect. We can never fully interact with BTBB reality, so, no matter how hard we try, we won't have the same satisfaction as when participating in physical reality. Theorizing about eternity being an instant offers little actual consolation. The tragedy of a young person's death lingers as long as ATB reality is so predominant. However, the closer we can approach pure BTBB interaction to help us understand the tragedy, the more intense our contemplation will be and the more relief we will feel.

In combating life's negatives, mankind has struggled to comprehend their underlying causes. Preachers have used the mysterious will of God to explain away tragedy, sickness, and failure. The existence of evil is blamed on the devil. Unbelievers don't agree, of course, and instead focus on physical causes, so they can improve whatever needs to be improved and prevent a repetition of the problem. Regardless of which viewpoint is adopted, there is a fundamental problem with any argument that doesn't take both realities into consideration. Whether one is a believer or nonbeliever, attempting to analyze total reality using only the criteria of one reality is a mistake.

Understanding the implications of BTBB interaction is particularly advantageous to believers who may question God's compassion on such occasions. A gnawing doubt may linger after such a tragedy: if God had wanted to prevent an untimely death, he could have done so. However, that's making a judgment based on an ATB interpretation of the events. That is, depicting God as uncaring ignores the possibility that God's time isn't the same as ours. An alternative conclusion—that God's different timeframe provides the explanation—may be harder to accept, and certainly its premise can't be proven, but it does offer another possibility.

What reality does God occupy? Is it the BTBB reality where our instant is its eternity, or could it be yet a third reality where its instant is our eternity? We don't know, of course, and can't know, but we can make an educated guess. If God's view of us was the same as our view of BTBB reality, then God's instant would be our eternity. That is, God's relationship to us would be equivalent to our relationship to the Big Bang. Yes, this is only a **guess**, but it is a reasonable one. Whether that guess is correct or not, however, the only way to make a judgment about God's motives using ATB reality is to assume that God is living within ATB reality. That would seem to be extremely unlikely.

Again, when we have a BTBB interaction, we interact with a reality having a vastly different speed of time. Billions and billions of years in BTBB's time make up our instant. As far as a God in yet another reality is concerned, that God's relationship with us could be similar to our relationship with BTBB reality. That is, all of our history would be God's **NOW**, or, in Latin, **NUNC**. Perhaps death and suffering need to be considered in the same light. What happens in our reality may appear to be cruel and heartless, but that may be because we experience it in slow motion from the viewpoint of that third reality.

Let's use the example of slow eating again. Say we make a rule that in eating a meal there must be seventy-five years between each chew. An average meal eaten under that rule would require millenniums to complete. It would take several lifetimes just to consume a forkful of corn. By the time the third chew was finished, 225 years would have elapsed. Talk about cruelty. Not only would such a meal be torture; the rule would be preposterous.

Yet that's exactly what we may be doing when we try to evaluate an Almighty's motives using our timeframe. Interaction with God involves interacting with multiple timeframes, and all of them have to be taken into account when making a judgment. Once that is understood, questions such as why it took billions of years for human beings to develop can be addressed from a much different perspective.

Of course, adding a third reality to our consideration is pure conjecture. This third reality may not exist, although I personally find it hard to believe that the two realities that earthbound humans experience are the only two. In addition, whether or not there is a third reality doesn't prove a caring God, or any God at all. But it does give us an objective perspective that makes the existence of a caring God more likely. Human life, with all of its turmoil and challenges, can be understood and accepted more easily once a consideration of multiple realities is included in the analysis. This more complete awareness of the human condition is yet another reason why knowledge of interaction with BTBB reality is so critical.

Those who believe in God may be comforted and supported by this concept. After all, it confirms their view that there is another reality out there, one that is nonmaterial and universal. Skeptics, on the other hand, may consider the idea a threat, since BTBB interaction seems to imply spirituality. However, neither view is necessarily correct. Remember, a true understanding of interaction with BTBB reality means that all reality is natural. That is, the spiritual **is as natural** as the material.

Chapter 32
Spirituality and BTBB Interaction

This book is not intended to challenge or critique efforts aimed at understanding the mystical aspects of human life. However, the properties of reality Before The Big Bang, timelessness, nonmateriality, and universality, are the same properties often involved in mysticism and religion. Thus, a discussion of interaction with BTBB reality inevitably raises the issues of mankind's search for an explanation of the inexplicable. In this chapter we will delve more deeply into the relationship between BTBB interaction and spirituality, keeping in mind that our definition of such interaction includes nothing mysterious. In the first part we'll see how the organizational aspects of formal religion tend to interfere with the potential for using religious practice as a means for understanding BTBB reality. In the second, we'll discuss how recognizing BTBB interaction helps us understand religious experience.

Using Organized Religion to Understand BTBB Interaction

We never attain pure interaction with BTBB reality; we can only approach it. We're embedded in the physical, and we even use ATB reality to facilitate what BTBB interaction we can achieve. That is, any BTBB activity, such as an artistic endeavor, requires material input of some sort which adds to the impurity of the BTBB interaction.

Arriving at a significant abstract conclusion is the closest thing to pure BTBB interaction we've discovered so far. A logical question would be whether religious experience is just as pure. The advantage of religion, of course, is that it is something that an individual can ponder at any time. That way a purer BTBB experience could be far more frequent and predicable.

Many people find comfort and truth in religious belief. The leaders of religious organizations are usually highly skilled, sincere individuals. Followers rightly respect their leaders' opinions and find satisfaction and solace in their advice. Practitioners are to be commended and respected for their commitment and devotion. In addition, both the followers and the leaders experience BTBB interaction through their faith, whether knowingly or not. For some, a religious service may be the closest they will ever come to pure BTBB interaction.

However, religious practice often includes at least the equivalent amount of ATB reality that is found in an activity such as artistic expression. The reason is that all organized religions, despite any claim of divine inspiration and intervention, are human institutions. They exist for human beings and are administered by them. Though usually experts in theological interpretation, the administrators thoroughly understand that we are embedded in ATB reality. Religious leaders, therefore, are often Doers with the same ambition and talents that entrepreneurs possess. These Doers keep their followers in line in order to propagate the message of their faith effectively. They are as necessary for religion as they are for any other human activity that exists in ATB reality. Doers keep the religion vibrant. Unfortunately, these Doer leaders can also unwittingly act as a barrier for understanding a religion's purer interaction with BTBB reality.

The analysis we make concerning organized religion is on a theoretical plane and should not be interpreted as a criticism of the leaders or followers of any particular belief. However, to find a purer form of BTBB interaction, we must look at the faith as outsiders, seeking to learn what those within the faith are already convinced they experience. We will see that many pitfalls are placed in the path of discovery and that overcoming these obstacles is necessary in order to find a high level of BTBB interaction.

Let's first look at typical leaders. Most religious administrators claim that we can can't experience pure spirituality in this world, and what spirituality we can attain is accomplished by living our ATB lives under their guidance. With that in mind, they often concentrate more on ATB than on BTBB reality, imposing a multitude of rules on how to exist in the physical world—what to eat, how to act, when to worship, etc. These same administrators work diligently at obtaining funds, so that the physical needs of the organization can be met. There's nothing wrong with this, of course, because otherwise the religion wouldn't survive. And, more important, their underlying assumption is right. Experiencing pure spirituality in this world would be like pure interaction with BTBB reality. It cannot occur. However, although an emphasis on ATB reality may be necessary for religious institutions, it restricts any inquiry such as ours, because it masks the actual BTBB interaction that faith can bring.

Since religious services and programs are typically entangled with ATB formality, a free, unstructured discussion with a sincere believer might be a more helpful way to learn about BTBB interaction within a faith. Of course, finding that believer will probably not be easy. Those most anxious to accommodate are often more ambitious to convert than to honestly share their spiritual experience.

Nevertheless, there are particularly devout individuals who use nonmateriality, universality, and timelessness as a ground for their spirituality, while at the same time maintaining sufficient interaction with ATB reality. They become models of holiness within their religion and are often recognized as such by the rest of society as well. Gandhi and Mother Teresa are prominent examples, but all faiths produce such individuals. However, the greater their reputation, the more likely it is that they will be personally inaccessible. If they've written about their experiences, reading their thoughts may be beneficial, but Doer administrators monitor that documentation closely. If there is a problem with an interpretation of ritual or belief, the writer must fall into line or risk ostracism. Thus, the enlightenment that this individual's purer interaction with BTBB reality should promote is tempered by the same Doer mentality that guarantees the continuity of the organization.

Delving into the religious theory instead of the organization itself could be an effective method for finding that purer interaction. However, the inquirer must first filter out all the extraneous rules and regulations that Doer leaders have inserted into the belief and which they fight to maintain. And again, pursuit of religious theory often butts up against the establishment which controls that theory's interpretation. If the organization considers a query about the faith as a challenge, there will be no cooperation. Even if the question is sufficiently deferential, there may be enough restrictions placed in the way to inhibit any breakthrough or discovery.

One way to get around this problem is to focus the inquiry on the religion's documentation. However, avid Doers don't make that easy either. Documentation is often so voluminous that separating the wheat from the chaff, the interaction with BTBB reality from the interaction with ATB reality, is time consuming and frustrating. There may be some contemplative books that will help, as long as they are not focused too narrowly or buried in mysticism. And, of course, the documentation won't refer to any other reality in the sense considered in this book. It will agree that another reality is out there, but that reality will be clouded in mystery and impossible to understand until the afterlife. If there is any interaction here and now, it can only be achieved through faith. Thus, the information may be there, but arriving at an accurate understanding of the religion's relationship to BTBB interaction will require considerable effort.

That doesn't mean that the attempt will be wasted. A dedicated pursuit may locate a willing believer or a writing that provides valuable insight into BTBB reality. While making the discovery may not be simple, the result may well be worth the effort.

Using BTBB Interaction to Understand Religious Experience

We have seen that religious organizations can make it difficult to use a study of faith to learn more about BTBB interaction. On the other hand, being aware of the existence of BTBB reality can make religious experience more understandable.

To demonstrate this, let's begin by noting that there is no correlation between religious belief and intelligence. There are highly intelligent believers and nonbelievers. And, if both are using correct logic, one would think that they would reach the same conclusion. Yet there is a contradiction—fact for one and fiction for the other—when the topic is the existence of a Prime Mover, the initiator of the Big Bang.

This fundamental difference appears almost impossible to reconcile; yet it isn't, provided the reconciliation includes both ATB **and** BTBB reality. To illustrate how, consider the following two examples of religious experience involving the question of God's existence. The examples will have the advantage of using actual individuals instead of fictional people. We will examine two women, both thoughtful and educated, and see how looking at reality affected their attitude towards religious belief. For anonymity, we won't use names.

The first woman had a strong faith and was married to a devout Christian. Both she and her husband had received a Christian education, attended church weekly, and raised their children in the faith. At one particular church visit, however, the wife suddenly became convinced that her faith was a sham, and she lost her belief in God instantaneously. While she still struggles to regain her former convictions, they have never fully returned.

The second woman was an atheist married to a man who shared her non-belief. The marriage worked with the common, fundamental acknowledgment that there was no God, and, while the misguided souls who believed in one might be sincere, they were also wrong. After a number of years, something equally spectacular but quite the opposite took place for this woman. She suddenly declared her belief in Christianity. Her husband was so astonished that he didn't know how to react. Eventually the marriage disintegrated, leaving the husband desperate to learn what had happened.

What's curious about both of these examples is that something very real and very tangible occurred that suddenly changed the way these two women looked at life. Here the protagonists were both feminine, but that fact has no bearing on the matter. The important thing is to discover what prompted these experiences. And why, whatever the cause, did it yield the opposite effect? We'll see how considering both ATB and BTBB reality provides a reasonable explanation.

We've already acknowledged that interaction with BTBB reality is personal and complex. Both of these women experienced an intense interaction. Let's look at those interactions from a total-reality viewpoint. In the first case, it appears that the wife realized that placing too much emphasis on BTBB interaction had prejudiced her outlook on ATB reality. She wouldn't have used those terms, but her thoughts were equivalent. Actually, the experience of "enlightenment" itself was a BTBB interaction. She interpreted it as a need to measure all her experiences against everyday, ATB reality. Those that didn't fit, she discarded. Since her faith, a purely spiritual interaction, was obviously not ATB, she put it aside. In the second case, the woman concluded that she had been making virtually all of her judgments based on ATB reality, and that reality didn't encompass her whole being. She knew there was more to life and decided to embrace a faith which satisfied that craving. For her, God became the ultimate source of BTBB interaction.

At the moment that these women had their experience, they overreacted, which is common. When we're convinced that we've been wrong, we often overcompensate. These women became so convinced that they had missed out on an important aspect of their life that they made up for it by overemphasizing interaction with the reality they had been avoiding.

The next step for these two women is to reconsider the experience. If they do so, taking into account **both** ATB and BTBB reality, they should be able to see the "conversion" in a different light. For the reformed Christian to ignore BTBB reality isn't any more reasonable than for the reformed atheist to ignore ATB reality. When they realize that that may be exactly what they're trying to do, they will become more open to finding a balanced outlook.

This discussion shows that recognizing BTBB reality within religious experience improves the understanding of what took place, making an objective analysis and a reasonable adjustment more probable. Further, firmly incorporating an understanding of BTBB interaction into life can enhance all religious experiences.

Using BTBB Interaction to Improve Communication Between Believers and Nonbelievers

So far, we have looked at spirituality and its relationship to BTBB reality in two different ways, trying to use religion to better understand BTBB reality and using BTBB reality to better understand religious experience. Now let's use BTBB reality to address another significant issue—the reconciliation of believers and nonbelievers. We may not get that second couple back together again, but we will see that understanding BTBB reality helps.

Believers and nonbelievers seldom communicate with each other on what both agree to be equal terms. To an atheist, the Big Bang was a natural event that required no assistance from a Prime Mover. Everything that's happened since has evolved naturally to form our present ATB reality—the only one that counts. The reality that atheists admit to is physical, whereas believers accept nonmateriality as a given. Time is looked at differently as well. For an atheist time is measurable, but, for a believer, spirituality has no time.

(This difference, incidentally, is mirrored in mankind's examination of material reality, and it could even be the reason for the two fundamentally different theories of how ATB reality works. First, there's the macro theory of General Relativity, and then the micro theory of Quantum Mechanics.)

Back to believer and nonbeliever communication.

To assure better communication, think of the possibilities if the atheist agrees that whatever existed BTBB may exist in some form in ATB reality. Of course, such a jump would be considerable, since preaching the spiritual to a skeptic is like communicating with a rock—there can't be a response. However, many nonbelievers frequently rethink their positions and are anxious for new insights on reality. By definition, however, any insight they accept must be one that's grounded in fact, not subjective opinion. So, for the sake of discussion, say some argument, such as ours, works. Most of the issues that concern the skeptic—differences in the definition of time and matter—will have been addressed. Thus, once the skeptic accepts the dual reality that all humans experience, a frank and honest discussion with a believer can commence. BTBB reality would explain the yearning for something beyond the physical that even skeptics possess, despite their conviction that there's nothing more to life than ninety years or so on this earth. If a nonbeliever accepts BTBB reality, therefore, it may offer him or her new paths of discovery, which could even progress to an acceptance of a reality beyond what humans experience during this lifetime. Even if that doesn't happen, at least the skeptic will have achieved a wider outlook.

Agreeing that humans interact with BTBB reality does not, in itself, prove the existence of God, but it does offer possibilities that could lead to such a conclusion. Of course, the fact that human thought migrates to the abstract may simply confirm that we're aware of our history. Just as we're inclined to live near water—a possible concession to the fact that life began in the sea—so too does mankind long to interact with the reality that existed Before The Big Bang. The existence of a Supreme Being is not necessarily coupled to that longing. On the other hand, reasonable people must ask themselves what BTBB reality actually is. If the only meaningful consequence of the "Bang" was ATB reality, why would BTBB reality play a role in our current existence? Pondering that question may be no more productive than counting angels on the head of a pin, but limiting BTBB's role to simply being the source of "The Bang" seems extremely restrictive. Why wouldn't it include a leftover that helps unlock the secrets humanity has long sought? A Prime Mover could easily be included in that picture.

Although BTBB interaction and spirituality are necessarily coupled, we've seen the difficulties formal religion presents in a search for purer BTBB interaction. Nevertheless, the coupling suggests a valuable opportunity for human progress in understanding religious experience. It also provides new possibilities for communication between those of differing beliefs, including those who actively disbelieve.

Spirituality provides hope for vast numbers of the population. That hope has necessarily been based on faith alone, since no purely physical explanation could be made for it. However, the recognition of BTBB interaction gives that faith-based hope some new credibility. Further, it offers fresh insight into the nonmaterial world to both the believer and the nonbeliever.

Chapter 33
The Ultimate Reunion

Having learned to look at reality from a much wider perspective, in this chapter we'll review what we've discovered, discuss its implications, and look for additional opportunities to apply our newfound understanding. We'll conclude by adventuring a bit further out and speculate about what may happen after death. That's risky, of course. Whether the speculation is oblivion, paradise, or some other outcome, it will be pure conjecture. Nevertheless, now that we have additional insight into reality, the subject is at least worth bringing up. What we propose may not be fully convincing, but its structure should be understandable, since it's based on the same concepts we've already been using. First, though, let's review the progress we've made.

Interaction with BTBB reality, though imperfect, does take place. BTBB reality exists within ourselves, built-in as part of the Big Bang legacy. We're children of that Big Bang, and, as such, inheriting its reality is a logical consequence. Thus, our interaction with BTBB reality is actual and just as fundamental to our being as the physical one that emerged ATB.

As ATB reality gives us feedback—a slip of the hammer gives us a sore finger; a properly designed ship floats—so too does BTBB reality, despite the imperfection of the interaction. An abstract conclusion that represents a breakthrough in human thought can be far more significant to a person's life than amassing wealth. A melody floating in the mind can be even more beautiful than one played on the piano. A stunning twist in a plot can be exciting to a reader, but its author may be even more amazed, wondering how the idea seemingly appeared from nowhere. We get feedback, just as we do with ATB reality, and that feedback is found within our thoughts. In BTBB interaction we find the nonmateriality, timelessness, and universality of another aspect of total reality.

Life teaches us many important lessons. At an early age we learn that we must eat and have shelter. Food satisfies the demands of hunger, and shelter satisfies our need for security. However, as we mature, we discover that we also crave the timelessness, nonmateriality, and universality that interaction with BTBB reality offers. BTBB interaction uplifts the spirit and promotes insights which lead to dramatic progress. Without interaction with both realities, a civilization will die. No food and there's starvation. No shelter and there's exposure. Without BTBB interaction, a society will stagnate and eventually lose its zest for survival. Society loses its human spirit, and a dull, drab existence replaces creativity and hope for the future.

Now we can appreciate why the abstract, the visual arts, humor, meditation, and music are so essential. Using the achievements of Albert Einstein's powerful intellect, we can understand what is actually happening and recognize how BTBB interaction differs from interacting with ATB reality. Moreover, since BTBB reality is the reality from which the ATB world is derived, we understand why there's a necessity to interact with both. We rejoice and bask in the satisfaction derived from pursuing abstract thought just as intensely as we celebrate a magnificent meal or a beautiful sunrise.

The extremists are an issue, of course. A Dreamer philosopher who places an undue emphasis on BTBB reality tends to ignore, or at least minimize, any interaction with ATB reality. That attitude can endanger the well-being of the philosopher as well as that of family and friends. Despite this risk, extreme Dreamers do play an important role in society, helping it understand how abstract reality works. Nevertheless, these Dreamers must realize that their preferred reality is only one aspect of the total. Doers have the same responsibility, of course. The Doer businessman who minimizes, even ridicules interaction with BTBB reality, may go unloved, even by his children. His wife could lose interest in him as well, even if she herself is a Doer. In any case, the marriage would be cold and hard. BTBB reality provides the human softness and empathy that we all crave.

The history of religion teaches us that we find our highest satisfaction when we balance our appetites for **both** ATB **and** BTBB interaction. Religions that attempt to stifle one of these appetites won't prosper unless their teachings change. Followers may tolerate the suppression for a while, but that acceptance can't last. Even the mentality of the Middle Ages—the belief that the purpose of human life was only to prepare for the hereafter—only lingered due to the fall of sophisticated Roman culture. Although it took centuries to fill in the void left by the collapse of Rome, the recovery eventually came. That recovery had to include accomplishment in both ATB and BTBB reality. Any view that demands interaction with only one of the realities cannot prevail.

Life holds many surprises and challenges, but more than anything else, life is opportunity. Each day offers the chance for discovery, joy, and accomplishment. What could be a better way to approach that opportunity than to pursue a goal that satisfies the totality of human need? Such a goal would recognize that there is more to life than the physical challenges that are so evident when we look in the mirror every morning. It would recognize those material aspects, of course, but it would also address the nonphysical and timeless needs that we all have.

An awareness of BTBB interaction offers an opportunity to better appreciate the goals of others as well. Think of those whom we previously thought of as lazy. Were they actually lazy? Or did they simply have a different "balance point" than our own? Perhaps they weren't as attuned to ATB reality as they should have been, but the term "lazy" implies incompetence, even indifference. Our new appreciation of human motivation teaches us that those we see as merely inactive are likely to be more in tune with BTBB reality. If they are lacking anything, it's an understanding that

ATB reality is just as important as the reality that seems to take up so much of their time. Remember, though, ATB reality is **just as**, not **more** important.

We've always known that life extends beyond the physical, but being aware of the two basic realities shows us that its non-physical aspect isn't mysterious or unexplainable. Time is relative, and mankind is embedded in a timeframe that's slower than the one that gave us our ATB reality. There is another, much faster timeframe, the BTBB timeframe, that has always called for our attention. Civilizations have attempted to interact with that other timeframe in a variety of ways. The ancients prayed to gods, played music, and told each other stories. They decorated caves with imaginative paintings. Think of it. Mankind has been drawing pictures of both ATB and imagined reality for all this time, but there has been no explanation why. Interaction with BTBB reality is why.

This book began by using the mathematics of a right triangle to demonstrate that time is relative. It's a hundred-year-old argument fundamental to the understanding of our total reality. The Theory of Special Relativity has been shown to be correct over and over again, and the implications of that concept form the basis for the thesis of this work. When two timeframes interact, a collision of perception is unavoidable.

If I climb aboard a rocket ship and head into space at a rate that approaches the speed of light, my time will be the same during the voyage as time is for the people I left behind. However, when I return, when I interact with earth again, I will find that my time was much, much slower during my voyage. However, it's **only** when I return that I experience the difference. Thus, it's only when we interact with BTBB reality that we perceive a drastic time difference. We spent a lot of effort searching for that perception, and we found it, albeit imperfectly, in thought, music, laughter, and meditation. We found it in its purest form, however, at the moment the human mind reaches a conclusion on a profound, abstract subject. The more profound the subject and the more revolutionary the conclusion, the closer the experience is to pure interaction with BTBB reality.

Of course, some may claim that BTBB interactions are not as important as interactions with ATB reality. After all, expertise in ATB reality is more rewarding for Doers than expertise in BTBB interaction is for Dreamers. Human experience, however, proves otherwise. While Doers receive economic rewards, Dreamers gain insight and develop new and exciting perspectives. Since BTBB interaction is personal, and therefore not necessarily applicable on the ATB level, its economic value may not be recognized. But Dreamers couldn't care less. Their reward is the experience itself. Being able to communicate that insight through the written word or through song or art may, however, offer financial gain and help the Dreamer survive in this ATB universe.

The personal aspect of BTBB interaction is arguably its most difficult feature. We all have an opinion as to whether a particular color or design is attractive, but the personal experience of BTBB interaction goes beyond mere taste. There is a fundamental difference in the way timelessness, nonmateriality, and universality play in each of our lives. Timelessness is particularly subjective. It doesn't mean that time is standing still, but that there is a difference in time when two realities that are at different speeds interact. The interaction itself defines the timelessness, and everyone has a personal experience in that regard.

Spacetime contains all ATB reality. ATB reality is locked into a system that can only be viewed in its entirety from the perspective of another reality—assuming one exists—that has the same relationship to us as we have with BTBB reality. That is, the events in our ATB reality could be witnessed at any time from a perspective on us that is equivalent to our perspective on BTBB reality. Remember, our ATB interactions don't happen and then go away. The light of a star may have left that star thousands of years ago, but it's just reaching the earth now. It's in our present, even though it's in the star's ancient history. As we see the light from our point in spacetime, our reality is defined for that moment, and then, for us, the light is gone. But it travels on, and eventually it will reach another galaxy and another point in spacetime where it will be as unique for a witness there as it was for us.

Ever since Einstein introduced his theory on time, people have fantasized about time travel but have never succeeded in making that fantasy real. The only way such travel could happen would be for us to find a way to interact with some other reality that's equivalent to the way we interact with BTBB reality. Imagine that we could find a viewpoint that has the same speed differential from our reality that we have traveling away from the Big Bang. Imagine also that we could interact with our present reality from that viewpoint. We would then experience our present reality similar to the way we now experience BTBB reality. Our new instant would be our present eternity. We would witness all of ATB reality in that instant.

One day physicists will discover the basic structure of the universe. They have been working on finding it since quantum mechanics identified the dichotomy between the macro and micro universes. When the solution finally comes, it's bound to be revolutionary. I suspect it will challenge an aspect of reality that, until then, we had all taken for granted—such as we did when we assumed that information could be communicated at infinite speed. Perhaps, as suggested in Chapter 22, the new breakthrough will originate in the way we look at dimension. Whatever it is, once it is identified, a new definition for reality will be established, just as when Einstein discovered the relativity of time.

Meanwhile, the rest of us will go about our business, combining interaction with the NOW and the BEFORE in the way we have always combined them. At least we have the advantage of knowing that we interact with two realities, and we have a definition of the more subtle of the two. Understanding interaction with BTBB reality allows us to give it its proper priority and to balance it with the more insistent reality that the ATB world presents. We owe it to ourselves and to our fellow man to understand and use timelessness, nonmateriality, and universality to the best of our ability.

How long has BTBB reality been around? For eternity? If it has, how long was that? How long is eternity? Remember, our time is much, much slower than BTBB's time. Only a fraction of a second of our time is an eternity in BTBB reality.

With that idea in mind, let's go out on a limb and try to make a reasonable estimate about what may happen after death. We don't know, of course, but it's interesting to wonder how time would enter into reality then.

Could after-death reality be the ATB timeframe? That would seem unlikely, since our physical bodies are embedded in that timeframe. Would it be the BTBB timeframe? That too would appear unlikely, since our time would be so fast that eternity would be only a moment of our present time. The more likely possibility is a third timeframe, one we speculated about earlier. If so, the relationship of that third timeframe to us would be similar to our relationship to the BTBB timeframe. **Our** eternity would be **its** instant. In death, we would experience the equivalence of time travel, after all. That is, our after-death clocks would suddenly run far slower than those on earth. Like the passengers on the rocket ship, we wouldn't notice the difference until we interacted with earth, but the difference would exist. Interaction with earth would occur when a loved one, such as a child, died. However, twenty four hours in the after-death timeframe would be millenniums in the earthly timeframe. So after death, in the new time, our offspring would arrive right after we do. There would be no waiting. In death we would not only be joining our ancestors, but moments later our children would join us. Then, almost immediately afterwards, our grandchildren would appear, then our great-grandchildren and our great-great-grandchildren. The rest of mankind would soon join us. It would be quite a reunion.

A lifetime of ninety years may appear long to us on earth, but to those in this third reality, that time is nothing. For that reality, earth's eternity is less than an instant!

Skeptics are probably chuckling at this point, and they could be right. Perhaps there isn't any afterlife. We've already admitted that talk of a third reality is pure conjecture, and the only hint that this third reality might exist is the interaction with BTBB reality. Those same skeptics might also insist that we don't interact with BTBB reality at all and would consider this entire treatise as nothing but fiction.

I challenge these skeptics to come up with their own explanation of where abstract thought and fantasy originate. Is abstract thought just some quirk of evolution that humans somehow found? Are revolutionary conclusions just a mystery that humans can't explain? Does only our "real" time exist? Is our choice to either ignore other, more profound explanations of life's complexity or simply accept those explanations on faith, unless or until they can be physically explained?

While the skeptics are working on that assignment, I'll be enjoying the idea of meeting my children and grandchildren right after I'm gone and bragging to them that I was right. Maybe to you readers as well. I hope I see you all.

Then, of course, the third reality would be confirmed. No surprise, though, right? After all, why should the ATB timeframe be the last one?

www.ingramcontent.com/pod-product-compliance
Lightning Source LLC
Chambersburg PA
CBHW030254290526
45785CB00001B/80